Springer Series in Advanced Manufacturing

For further volumes:
http://www.springer.com/series/7113

Federico Rotini · Yuri Borgianni
Gaetano Cascini

Re-engineering of Products and Processes

How to Achieve Global Success in the Changing Marketplace

 Springer

Federico Rotini
Dipartimento di Meccanica e Tecnologie
 Industriali
Università di Firenze
via di Santa Marta 3
50139 Florence
Italy

Yuri Borgianni
Dipartimento di Meccanica e Tecnologie
 Industriali
Università di Firenze
via di Santa Marta 3
50139 Florence
Italy

Gaetano Cascini
Dipartimento di Meccanica
Politecnico di Milano
via Giuseppe La Masa 1
20156 Milano
Italy

ISSN 1860-5168
ISBN 978-1-4471-5798-4
DOI 10.1007/978-1-4471-4017-7
Springer London Heidelberg New York Dordrecht

ISBN 978-1-4471-4017-7 (eBook)

Printed on acid-free paper

Springer is part of Springer Science+Business Media (www.springer.com)

Preface

An abundant amount of literature regards the twenty first century as the innovation era, where shared knowledge fosters the progress with a strong impact at the economic and social levels. As a consequence, the capability to innovate will play a growing role in the destiny of companies. In the transition between quality-oriented and innovation-driven competition, considerable difficulties are faced by those firms, whose management mindsets have not assumed knowledge and innovation as a central asset of industries. With reference to such problems much academic research has been carried out in order to deal with continuous innovation programs, but the proposed indications have not resulted yet sufficient to master the competition, avoiding waste of resources and wrong decisions.

The debate about innovation and especially the factors that determine the success of innovation initiatives is rich and thought provoking. Scholars from different domains, especially business management, engineering, and computer science have disputed about the priority of the measures to be attained to achieve effective innovation tasks. Several models and strategies are grounded in concrete evidences from successful experiences. Is any proposed model capable to describe and justify any kind of successful innovation initiative? The authors' answer is "no" and we are convinced to share this opinion with the wide majority of industrial experts, researchers and readers. We believe that no tool has been developed to support the whole innovation cycle and to cover any aspect (strategic, technical, managerial, etc.) regarding industrial approaches to innovation.

Undoubtedly, this book cannot fill the gap. The issues that are treated to infer the motivations of the work are surely not exhaustive in the panorama of innovation strategies. Much academic research and industrial practice is still needed to build rigorous models and efficient tools. However, we are convinced that such a manuscript and the presented techniques can result in a useful support to industries facing the need to undertake decisions about the renovation of processes and products and a valuable contribution to researchers and PhD students who are interested in the field.

The whole coverage of this book swivels on a basic assumption, that results unopposed in the literature and that we have considered as a fundament for the book: the capability to provide customer value is a primary driver in achieving business success. According to our studies and background, such hypothesis can be deemed valid in both static and turbulent stages that characterize the paths of evolution encountered by goods and services. If the focus on value contributes to sustain successful business, all innovation initiatives, being addressed at products or processes and providing radical or marginal changes, cannot overlook their potential impact on customer satisfaction. In such a perspective, the combined set of presented value-oriented methodologies, namely Integrated Product and Process Reengineering (IPPR), constitute the core of a toolkit for the identification of the most favorable directions within innovation initiatives. IPPR represents a system to support crucial decisions in the industry, capable to orientate choices among a set of plausible reengineering activities, according to value criteria. The methods that we illustrate in the present publication deal with different specific objectives and conditions encountered along industrial production, research, and planning. Actually, three recurring situations are taken into account, that typically take place along product lifecycle: the birth of new products and the organization of suitable processes, their maturity often accompanied by lacks of competitiveness, the need to drastically rethink the outputs that are offered to customers.

Acknowledgments

The authors acknowledge the aid provided by anonymous reviewers, contributing to the drafting of the present volume, as well as to the writing of previous manuscripts treating matching themes. They really shed light on the potential of the developed instruments, proposing further developments that have been partially addressed within the present coverage.

The authors are indebted to some of their students, employing the presented techniques for experiments on which they based their MS thesis and all their colleagues, who provided suggestions about the treated subjects and contributed to the development and testing of the proposed tools. For this purpose we would like to mention the support provided by Daniele Bacciotti, Alessalndro Baldussu, Niccolò Becattini, Alessandro Cardillo, Walter D'Anna, Lorenzo Fiorineschi, Francesco Saverio Frillici, Luca Lazzarini, Massimo Lotti, Fabio Piccioli, Francesco Pucillo. The encouragements by Professors Umberto Cugini and Paolo Rissone to investigate the treated arguments and write the present book resulted in not minor sources of motivation.

Contents

Chapter 1
Introduction

1.1 Generalities of the Reengineering Strategies

By treating the problems related to products and processes, thus the outputs of industrial activities and the ways to achieve them, a large range of business, technical and organizational features have to be taken into account.

The present Chapter first overviews general and comprehensive models that attempt to deal with a large range of the recalled aspects and provides the relevant definitions for the main concepts encountered in the book. According to this aim, this Section illustrates the main notions concerning Business Process Reengineering (BPR) and New Product Development (NPD), in order to facilitate the reading and understanding of the investigated topics. The field of BPR includes a broad set of techniques to approach the task of improving the internal processes, by optimizing the allocation of resources, the employment of skills, the performances of the end products, etc. Conversely, NPD groups the most common practices undertaken to innovate products and services.

Section 1.2 is dedicated to point out the product lifecycle phases that are addressed and supported by the instruments described in the book. A more detailed review of the subjects of interest is presented in Sect. 1.3, that eventually points out the main identified deficiencies and open issues with reference to the surveyed methodologies and models. Finally, Sect. 1.4 clarifies the purpose of the book, addressing the general goals and the methodological objectives to be attained by building a versatile system to support decisions in troublesome industrial contexts.

1.1.1 Redesigning Business Processes

The concept of "*business process*" was born in the early 1990s as a means to identify all the activities that a company performs in order to deliver products or services to their customers. The need of describing and formalizing the actions

F. Rotini et al., *Re-engineering of Products and Processes*,
Springer Series in Advanced Manufacturing, DOI: 10.1007/978-1-4471-4017-7_1,
© Springer-Verlag London 2012

performed to turn resources into benefits for the customer was strongly perceived in those years since companies started worldwide to radically reorganize their activities in the attempt to regain the competitiveness lost during the previous decade. The "business process" concept has been defined by several authors in the literature with the aim of providing a reference for modelling and analysis tasks.

Davenport [1] stated that it is "*a structured, measured set of activities designed to produce a specific output for a particular customer or market. It implies a strong emphasis on how work is done within an organization, in contrast to a product focus's emphasis on what. A process is thus a specific ordering of work activities across time and space, with a beginning and an end, and clearly defined inputs and outputs…. Processes are the structure by which an organization does what is necessary to produce value for its customers*". Thus, according to Davenport, a business process is identified through clear boundaries, inputs, outputs and activities ordered in time and space: the purpose of the process is the transformation of inputs into outcomes having value for the customer.

Hammer and Champy [2] give a more general definition focused on the process outcomes according to the customer perspective: "*a collection of activities that takes one or more kinds of input and creates an output that is of value to the customer*".

Eventually Johansson et al. [3] emphasizes on the creation of links and inter-relations among the activities and on the transformation that takes place within the process, highlighting the value chain concept: "*a set of linked activities that take an input and transform it to create an output. Ideally, the transformation that occurs in the process should add value to the input and create an output that is more useful and effective to the recipient either upstream or downstream*".

Plenty of definitions have been proposed, but in essence all have the same meaning: business processes are basically relationships between inputs and outputs, where inputs are transformed into outputs throughout a series of activities, which add value to the inputs.

According to the cited contributions, a business process should be therefore characterized by (Fig. 1.1):

- clearly defined boundaries, inputs and outputs;
- activities ordered in time and space;
- a clearly identified beneficiary of the process outcomes, e.g. the customer or any stakeholder;
- the transformation taking place within the process that is meant to add value to the inputs;
- an organizational structure;
- one or more functions to be performed.

Such properties suggest that the business process can be considered as a technical system able to generate value by manufacturing products or delivering services under certain boundary conditions such as market demand, raw material availability, product requirements, technology and know-how resources, etc. When the process is not able to exploit the available resources according to their

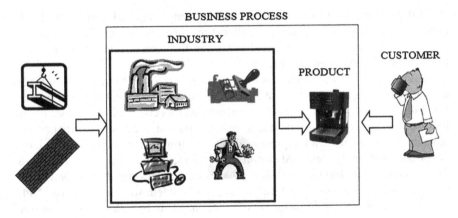

Fig. 1.1 Elements constituting a business process

potentialities, its capability to survive market competition decreases dramatically, due to a disadvantageous balance between the provided benefits and the involved costs. Thus any organization has to pursue continuous business improvements through planned evolutionary paths in order to preserve its competitiveness; this evolution can involve the business at different levels and it requires resources of knowledge dispersed across different fields and disciplines.

During the last twenty years, many methods have been suggested to address the redesigning and innovation of business processes. In the management field, but also in the scientific literature, such reorganization tasks were grouped under the name of Business Process Reengineering (BPR) activities. Several definitions of BPR are available but one of the most acknowledged is that provided by Hammer and Champy [2], who depict it as *"the fundamental rethinking and radical redesign of a business process to achieve dramatic improvements in critical contemporary measures of performance, such as cost, quality, service, and speed"*.

In line with the above definition, IPPR provides appropriate hints to direct the process reengineering efforts towards solutions that comply with products quality, customer satisfaction, resources savings and suitable sequences of phases to achieve such issues.

1.1.2 Rethinking Products and Business Models

In particular circumstances, it is required to radically redefine the outputs of the business process, rather than to rethink the ways to generate them; as a result, the core of the innovation task shifts towards redesigning, manufacturing and delivering new products. In this context, NPD cycles are typically concerned as conjoint activities involving business, marketing and technical expertise within companies. The continuous information exchange among the various units of the firm debates about the feasibility of new products, the knowledge required during the development stage, programs for resources allocation, the financial sustainability of the project, the target group of expected users, the foreseen response from the customer arena, the commercial strategy, etc.

Besides, it is acknowledged how the destiny of NPD initiatives is mainly determined during the Fuzzy Front End of the design cycle, i.e. the initial phases, when the design process appoints the fundamental aspects of the new goods [4–6]. In such segment of the development task the human needs to be satisfied are crucially individuated and represent the main trigger for exploiting business opportunities. Clearly identified customer requirements are regarded as the inputs for the conceptual design, whereas product specifications are defined and eventually conflicting demands are faced throughout problem solving tools. The NPD cycle is therefore grounded on the set of requirements to be fulfilled, which in turn provide value for the target customers. The beginning of the development process captures plenty of the characteristics related to the final commercial offer, ranging from the identity of the company which strives to address unspoken needs to technical performances to be achieved in order to cope with established exigencies. In this sense the NPD task commences by giving prominence to the ways to manifest the culture of the company, the position it is ought to be gained in the marketplace, novel paths intended to deliver value to customers and thus, altogether, what we can identify with the concept of *"business model"*.

From a historical perspective, the wide diffusion of the "business model" term is consistent with the growing role played by Internet and particularly by the e-commerce in marketing activities. By the 1990s of the previous century, the adoption of web retailing was considered as a sort of mantra for determining companies' fortune. Despite such enthusiasm, numerous e-commerce experiences resulted in tremendous flops, as surveyed by Mahajan, Srinivasan and Wind [7], because of their lack of strategy within flawed business models [8]. As a consequence, the notion of business model started to assume a wider meaning and to identify patterns of value creation by exploiting business opportunities [9]. On the same wavelength Chesbrough and Rosenbloom [10] individuate the primary objective of the business model in the proposition of the value necessary to provide commercial interest to technological advances. In more general terms Francis and Bessant [11] address the objectives of *business model innovation* in the *"reframing of the current product/service"*, thus allowing to individuate "new challenges and opportunities". In order to fulfil the task, Johnson et al. [12] depict Customer Value Proposition as the first step in the creation of an alternative business model with the aim of fulfilling unsatisfied needs.

The efforts to redesign business models, and consequently NPD tasks, are therefore associated with value innovation initiatives. Value innovation is acknowledged, also within studies about entrepreneurship, as a fundamental strategy to obtain competitive advantage by proposing value profiles that deviate from previous industry standards. The renewal of business models is intended as a means to achieve differentiation from competitors in ways valued by market [13, 14], assuming distinguishing features with respect to any other sort of disruptive innovations [15]. Indeed, such kind of innovations fundamentally redefines the market boundaries through New Value Proposition (NVP) initiatives that emphasize on previously overlooked product or service attributes, which result valuable for the customers. In this perspective Gotzsch, Channaron and Birchall [16] emphasize the value provided by communicative capabilities of products, especially when the main features, e.g. performance and price, have reached their maturity.

In order to clarify the meaning of the introduced terminology, within the present book we depict a *value profile* (or *value curve*, with the reference to the graphical model introduced by Kim and Mauborgne [17]) as a *bundle of properties and features, characterized by their performance or offering level, belonging to a product (or service), which generate benefits for the user*. Such features are indicated throughout the text with attributes (as observed through the lenses of end users) or customer requirements (from the viewpoint of the enterprise), whereas the *product* they belong to, is intended as *the output of the business process, thus a set of tangible items and matched delivered services*. Moreover we exploit the definition provided by Barnes, Blake and Pinder [18] who identify a value proposition as *"a clear, compelling and credible expression of the experience that a customer will receive from a supplier's measurably value-creating offering"*. As a result NVP pursues the objective of differentiating value profiles from those existing in the industry, with the attempt of developing new generations of products and services that enhance customer satisfaction by offering in a synergic way additional benefits and unprecedented experiences.

1.2 Classes of Reengineering Problems

Facing innovation issues, companies typically have to pursue the double goal of delivering customer satisfaction and carrying out industrial processes with a limited amount of expenditures and consumed resources. This implies that the development of products and processes involves tangled interrelations between companies policies and the features affecting the market and the customer perception.

In a simplified vision, as already advanced by Miles' Value Engineering [19], the firms have to maximize the ratio between the profitability of the delivered products and services and the costs pertaining the business process. By considering the profitability directly related with customer satisfaction, quality oriented tools have been developed to increase the numerator of the ratio. A wide diffusion in the industry regards the *Quality Function Deployment* [20], a mathematical model to maximize customer satisfaction by individuating the most advantageous combination of product attributes performance within the range of feasible technical solutions. On the other hand, the most diffused BPR strategies are aimed at shrinking the production costs (thus minimizing the denominator), by eliminating all the superfluous expenditures and thus obtaining lean processes. In such framework, most of the work has been dedicated to optimize the terms of the fraction (benefits and costs), rather than the ratio as a whole. Within the landscape of reengineering techniques, the

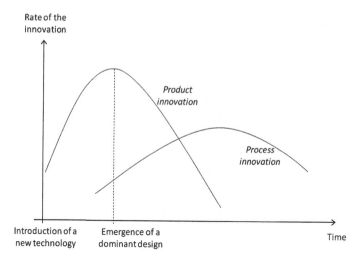

Fig. 1.2 Innovation timing according to the dominant design model

present book provides a contribution in terms of their capability of taking into account the aspects related to both processes and their outputs.

However, the various tools introduced through IPPR are consistently either process-oriented or product value-driven. The choice is dictated by the evolution of industrial products, which alternates the predominant relevance of performances and costs. Well-established models, such as the one developed by Utterback and Abernathy [21] highlight, as depicted in Fig. 1.2, that the focus of industrial efforts shifts along systems lifecycle from product characteristics to processes (and vice versa): more specifically the breakpoints are represented by the emergence of a dominant design and the introduction of new technologies. Thus, firms should prioritize their reengineering endeavours according to product development stages and market conditions. On the basis of such evidences three different classes of industrial problems are outlined and described as follows. As a result of sustained innovation, it is likely that the tools introduced to support the classes of problems are to be used in a cyclical fashion, e.g. the definition of new product features determines the need to reconsider the value impact of the phases belonging to a prototypal industrial process.

1.2.1 Class of Problems #1: Organize a New Process to Overcome Market Boundaries

According to Pahl and Beitz [22], the need to develop new products or services arises from different stimuli which may be internal or external to an organization. They can be summarized as follows:

- the market: new customer needs to be fulfilled, new functionalities to be delivered, etc.;
- the company: new ideas and results coming from research and development activities, availability of new manufacturing technologies, etc.;
- other: new policies, environmental issues, etc.

Typically such stimuli represent great potential inputs for individuating new business opportunities, as described more in detail in Sect. 1.2.3.

However, the exploitation of the previously individuated stimuli may result in the inadequacy of the know-how, the technologies or the managerial skills belonging to the firm in terms of delivering the new elements of value. Consequently, novel product ideas often encounter significant problems to access the market due to a large amount of factors such as design or manufacturing costs [23], organizational issues [24], required technologies or materials [25], relevant drawbacks [26], resources consumption [27].

In other circumstances, changes in the boundary conditions, such as shortage of materials or semifinished products, consistent modifications occurred along the supply chain, rise of costs, come out as relevant hindrances to business.

In all the above described conditions, the necessity emerges to basically rethink the business process, due to the insufficient generation of value or profit drops.

Preliminary hypotheses about how to overcome the mentioned under capacities may result in unsatisfactory processes and/or limited performances. The tools developed within IPPR to address such kind of problems are aimed at evaluating which phases and activities require major adjustments in order to align the business process to the expected delivery of value.

1.2.2 Class of Problems #2: Individuate the Bottlenecks that Generate the Loss of Competitiveness

Also after their launch, all the products have to pursue continuous improvements in order to satisfy changing customer requirements or market conditions. This task implies an evolution of production and business processes at different levels. In some circumstances minor reorganizations in the design or production phases can be sufficient to fulfil the evolving demands; besides, these actions usually bring only to limited improvements, mainly focused on preserving the appeal of the product in the marketplace. In other market circumstances, as in event of declining competitiveness, the companies have to develop more remarkable innovations.

In order to envisage an enhanced business process addressed at gaining competitiveness, the current sequence of industrial activities can be analyzed to highlight strengths and weaknesses. Thus, investigating the process with a focus on the value provided to the end user means analyzing the contribution of each phase to the generation of product/service attributes. Such task has to take into account the impact of the phases in determining the customer satisfaction, considering as well the involved resources in delivering such contribution. As a result of the analysis, it may happen that one or more phases of the process provide a marginal contribution, show poor performances or their fulfilment determines an excessive consumption of resources. On these bases, the main value bottlenecks are individuated and the suitable reorganization actions can be prioritized to overcome the emerging deficiencies. Subsequently, the process can be reengineered by identifying the proper technical solutions that implement the individuated actions.

1.2.3 Class of Problems #3: Build the Value Profiles of Innovative Products

When facing a NPD task, the common approach is to reach a trade-off among the established and new (if any) parameters or features that characterize the redesigned system. Such reengineering mode follows a cautious logic of business transformation and typically brings to minor improvements, disregarding the early intentions and missing to exploit the initial ideas. Although a not marginal school of thought encourages companies in preserving to compete with their established business models by delivering incremental improvements, a growing amount of researchers supports the need to perform more radical innovations, as in [28]. To this end, a process-oriented approach is not capable to effectively address reengineering activities focused on radical product innovation. The need to carry out more disruptive innovations takes place especially when established technologies get superseded or customers start to attribute value to novel product features.

With reference to the first case, according to a well-established model, the main performance of a system grows by typically following a S-shaped curve [29] as a function of the research effort that has been dedicated to its development. When the system has reached its maturity stage, its evolution approximates a limit with hardly appreciable improvements. In this phase, the industry gradually adopts emerging new technologies, which are capable to overcome the previous performance limits of the system. The phenomenon is graphically depicted, as shown in Fig. 1.3, through the birth of a novel S-curve, which gradually grows, hence surpasses the old performances standing still and supersedes the preceding technology. Such representation is commonly employed to illustrate long periods of incremental development of the systems and turbulent phases characterized by technological ferment and thus radical innovation.

Besides the "natural" pattern of growth, the shift to technologies with higher performances can be dictated by external factors. Typically, remarkable discontinuities in customer demands and preferences lead to the need to perform technological advances, implying to acquire wider knowledge to stay ahead in the competition.

In the cases regarded as radical innovations driven by value issues, the innovation process has to fulfil needs previously unsatisfied within a given industry. Considerable advantages can arise by investigating further aspects of value that consumers might care about, such as way of using, overlooked technical or emotional features, resources committed to the user, maintenance, environmental

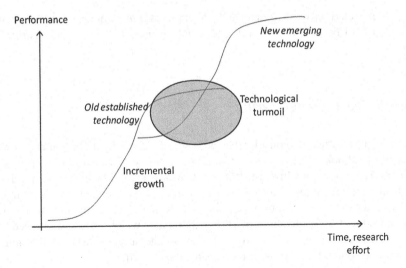

Fig. 1.3 S-curve model of technological substitution

impact, customer care, end of product lifecycle. Additional attractive features can be thus attributed to the current product profile with the aim of boosting customer satisfaction. The focus on customers and on their perception of products and services has gained an increasing role within companies, by affecting industrial practices, policies and product development cycles, leading moreover to a great impact on the required managerial skills.

In each case a difficult task is yet represented by the implementation of suitable technical and organizational solutions which should allow to perform the manufacturing and the marketing of radical innovations. Such phase is commonly characterized by the emergence of conflicting requirements and the involvement of knowledge dispersed across many disciplines. Given the great consequences of the success or the failure resulting of such an overwhelming design effort, suitable supports for directing radical innovation processes would result considerably valuable.

According to Christensen's studies [30, 31], the first mover advantage is particularly relevant when the introduced products are characterized by additional features, rather than new technologies. On the same wavelength, as already remarked in Sect. 1.2, a thread of business literature emphasizes how value differentiation can result in profitable innovation experiences. Given the consistent advantages attainable through NVP strategies, useful tools have been developed within IPPR to support companies in performing this sort of initiatives.

1.3 Brief Review of Tools and Methods Available in Literature

As already recalled, industries face the challenge of suitably and continuously improving their offer in terms of marketed products and delivered services, as required by rapid changing and highly competitive marketplaces. In order to fulfil the new expectations of customers and stakeholders, innovation and reengineering efforts have to be oriented towards industrial business processes and distributed products.

1.3.1 Process Reengineering

The literature witnesses considerable advantages arisen by BPR initiatives and describes textbook success stories. Conversely, plenty of contributions from different periods point out a high percentage of unsatisfactory results concerning BPR practical implementations, causing therefore diffused scepticism in the field. In the 1990s of the previous century, Holland and Kumar [32] estimated in a range between 60 and 80% the share of BPR experiences which have not pursued the expected out-comes. More recently, a review of the success rates of BPR initiatives has substantially confirmed those percentages [33].

Beyond the excessive expectations placed by managers and CEOs, the reasons of diffused BPR flops can be related, according to literature, to three main motivations: the diffused disregard of the appeal of process outputs, the complexity of industrial systems to be administered, associated with uncertainty issues, and the defiance of people in the organization.

1.3.1.1 BPR Deficiencies in Focusing on Success Determinants

BPR strategies strive to take into account a wide range of features relevant to the industries, such as price, lead-time, delivery conformance, performance, quality and reliability, sources of risk, environmental factors and life-cycle costs. Nevertheless, most of the developed approaches have been characterized by the priorities assigned to one or more of the previously listed features, meant as the main triggers to successfully perform the product and process development [34, 35]. Furthermore, under the aegis of BPR, countless projects have been carried out to promote the introduction of Information and Communication Technologies (ICT) [36] and most of them reflected the intended aim of shifting towards Lean Manufacturing practices [37].

Herron and Braiden [38] have developed a methodology to assist the user in identifying the most appropriate lean manufacturing tools and techniques to address the problems of a particular company through a quantitative compatibility assessment. The results confirm that lean manufacturing tools may have a major impact only on specific areas of the business, but they are not a panacea for any kind of problems. Typically, companies experienced problems in areas such as under capacities, scheduling and innovation in products and processes, which represent issues that are not directly influenced by lean manufacturing methods.

In such framework, failures of BPR initiatives can be explained by strategies oriented on redesigning just the features pertaining the internal processes [39] and focused mainly on resources savings [40]. A great quantity of experiences, with the objective of achieving lean processes by mimicking past experiences, have frequently underestimated the relevance of the value delivered to customers [41], seen as a determinant for the success of BPR initiatives [42]. The BPR goal of building and organizing a business architecture aligned to fulfil customer demands, as depicted by Edwards and Peppard [43], has thus often been disregarded.

From this point of view, other methodologies have been proposed in order to batten down the hatches, by proposing approaches that take into account additional objectives than costs lowering.

A not negligible amount of works approaches the problem of dealing with concurrent issues in terms of costs management and product requirements; an example is [44], where the integration of Value Engineering and Target-costing techniques is proposed to support the product development process in an automotive company. Such a methodology was applied to a case study aimed at improving costs and performances of a vehicle engine-starter system, according to customer and company needs. In [45], an integrated multidimensional process improvement methodology has been proposed to address the yield management, process control and cost management issues for a production process. Total Quality Management (TQM) is used to manage the cost of the system according to the quality requirements and a discrete event simulation is employed to achieve process reengineering and improvement. Another method has been presented in [46] based on a heuristic approach which supports the practitioners in developing a new improved business process starting from the current design. The method has been extrapolated from different successful and acknowledged best practices to carry out BPR tasks. These heuristics have been synthesized in a checklist for process redesigning with the objective of contemplating and harmonizing different management approaches: Total Cycle Time compression, Lean Enterprise and Constraints Management.

In the panorama of the techniques supporting business management, the Balanced Score Card (BSC) [47] represents the most established approach to identify reengineering directions on the basis of multiple criteria. Its main strength stands in combining financial and nonfinancial performance indicators in a coherent measurement system. The enterprise is evaluated according to indicators belonging to four different areas: the financial perspective; the customer satisfaction; the internal business process view based on the concept of the value chain; a final index taking into account the innovation and the learning perspective. The advocated deficiencies of the strategy regard BSC limitations as a result of invalid assumptions within the innovation economy [48]: its rigidity, its conception of knowledge and innovation as a routine process, its focus on the internal processes of the company determine biased evaluations since the relationships with the environment are neglected. Such limitations make the BSC performances poor in event of radical business modifications that frequently come out in the innovation age.

1.3.1.2 BPR as a Support for Decision Making and Decision Support Systems for BPR

From a methodological point of view, BPR applications represent complex multidisciplinary tasks, dealing with multiple sources of risk [49] and a wide range of facets regarding different fields of expertise [50]. On the same wavelength, Ozcelik [33] underlines how the major risks related to BPR implementation regard projects involving different functional units of the companies.

Furthermore, reengineering issues have to be directed towards complex systems, such as business and industrial processes, which have by nature not deterministic behaviours [51] and that require dynamic time-dependant models. As a consequence, the uncertainty regarding the model and the parameters governing the business process affects the outputs of BPR tasks, leading firms to take extremely risky decisions in order to pursue the planned enhancement strategy. It follows that the development of Decision Support Systems (DSSs) aimed at addressing the most appropriate directions for redesigning industrial processes represents a flourishing research field.

Like each decision-making activity, the redesign and planning of business processes is associated with uncertain inputs and risk. Lambert et al. [52] take into account relevant risky factors starting from the modelling phase by representing such additional information in IDEF frameworks.

Many research efforts about DSSs dealing with the uncertainty that characterizes a business process have been carried out; their complementary aims range from enhancing specific aspects of the industrial strategy, to supporting the development of certain categories of firms and increasing well identified performances. Min et al. [53] developed a decision support system suitable for banking industry, assessing appropriate Business Process Reengineering tasks under multicriteria analysis and present constraints. Williams et al. [54] deal with risk and uncertainties associated with BPR initiatives, focusing on organizational hurdles and providing guidelines for pursuing incremental or radical changes with reference to expected benefits and available investments. Wang and Lin [55] introduced genetic algorithms in order to efficiently schedule industrial processes for a make-to-order manufacturing firm. Their research and application is tailored for resource allocation decisions in an environment characterized by time pressure with regards to delivery dates. By exploiting simulation techniques Mahdavi [56] built a model meant to dynamically control the production activities of a flexible job-shop, whereas manufacturing processes are characterized by stochastic events.

Another branch of the research that involves intelligent decision making within industrial processes affected by uncertainty regards methods tailored for choosing the most favourable alternative among a set of already identified opportunities. Through the employment of simulation models addressed at treating uncertain inputs, Völkner and Werners [57] developed a decision-support system for choosing the best option among alternative business processes, approaching the problem with quantitative parameters. The developed tool is consistently tailored for those cases involving decisions about operations sequences. Gregoriades and Sutcliffe [58] proposed a decision-based system, taking into account industrial performance and human factors, capable to evaluate the advantages of introducing and managing a new candidate business process. The system simulates the business process and assesses further opportunities and risks, providing statistical outputs with reference to the generated scenarios.

Still with reference to BPR tasks, the problem of working with not deterministic and uncertain models is compounded by the presence of qualitative parameters. In such framework recent contributions introduce measurable parameters to deal with uncertainty issues within relevant aspects related to business processes, i.e. customer relationship [59] and purchasing management [60]. In order to compute even qualitative aspects, He et al. [61] have developed a Fuzzy Analytical Hierarchy Process to support the choice among different BPR alternatives.

Still within the development of tools to support decision making, Ramirez et al. [62] point out how business process redesign initiatives and conjoined IT advances positively affect product performances. Their research sheds light on the opportunities provided by process-oriented programs in order to achieve managerial success, regardless the implementation of novel IT supports. Therefore the analysis of industrial processes and its implications represents a relevant starting point for designing business advances.

1.3.1.3 Social Fallouts of BPR Experiences

Although our work tries to contextualize BPR in the field of engineering innovation, we cannot neglect the most frequent and harsh critique which gets directed towards most of these applications. The literature charges BPR about its strict focus on efficiency and technology and the disregard of people involved in the initiatives: not just customers and stakeholders, but also the labour. Very often the label BPR was used as a justification for major workforce shakeouts with the aim of decreasing organizational and production costs, instead of suggesting any kind of improvement based on process innovation. Knights and Wilmott [40] overview numerous contributions that demonstrate job losses (as supported by Grover [63]), complaints of workers subjected to BPR experiences, scarce consideration of labour welfare. Grint [64] goes as far as to talk about alienation amongst workforce. Additionally, whereas BPR applications have turned out as a misfortune for the employees, on the other hand the disregard of human factors have hindered the successful display of reengineering programs, thus resulting in complete disasters.

Management practices, such as those regarding human resources, do not fall within the objectives of the present book. However, it is useful to remark that a full understanding of successful reengineering projects advocates dealing with, among a wide range of factors concerning management disciplines [65], human factors and employment issues without making the innovation initiative a "blood, sweat and tears" experience for people.

1.3.2 Product Reengineering

As recalled in Sects. 1.1 and 1.2, several studies demonstrate how innovations in terms of the value directly perceived by the customers deliver the firms fundamental benefits in terms of market success. The literature includes contributions tailored to maximize the customer satisfaction and proposals intended to exploit unprecedented sources of value.

The methods of the first set (Sect. 1.3.2.1) are intended to design products whose mix of characteristics and performances ought to attain the greatest appeal of customers within the range of known technical solutions. Such techniques are consistently based on people opinions and preferences expressed through market researches, whereas the extracted data are used as inputs for optimization procedures. These approaches concentrate just on the explicit and revealed needs to be fulfilled, yet they usually do not bring to the identification of unexplored business opportunities or to disruptive innovations.

A particular set of tools belonging to the second group makes reference to the tendency of integrating manufactured products and a bundle of associated services (Sect. 1.3.2.2). Other methods are addressed at stimulating the generation of new business ideas swivelling on the fulfilment of different, diffusely latent, customer needs (Sect. 1.3.2.3). With reference to this kind of methodologies, the actions directed towards the fulfilment of superior value are mostly aimed at breaking quality/cost tradeoffs within the established competition in the reference industries, thus putting in practice tremendous differentiation strategies.

1.3.2.1 Innovations Based on the Optimization of Product Performances

In order to comprehensively pursue the fulfilment of the product requirements elicited by users, several methods have been developed in the consumer research field, with the objective of capturing the so called "Voice of Customer" (VOC); in [66] an extensive survey is presented. Many approaches such as those based on Free Elicitation, Laddering, Conjoint Analysis, etc., try to assess the product attributes having major interests for the user by interviewing techniques in which the customers are asked to identify the characteristics they consider relevant in the perception of a product. Other methodologies (i.e., Empathic Design, Information Acceleration, etc.) are based on observing the consumer behaviour during the day life. The assumption behind these approaches is that designers can easily identify opportunities for products in response to perceived needs, by examining the consumer behaviour. Without shedding light on novel and potentially valuable attributes, as claimed by Ulwick [67], asking the customers helps just to reveal the needs they are clearly aware, since they are capable to figure out only feasible solutions regarded of minor product improvements. Bower and Christensen [68] go as far as to claim that the inability of firms to efficiently innovate is caused by the aptitude in strictly meeting current customers expectations.

Along the product development process, once the main attributes of value have been established, the application of Quality Function Deployment (QFD) is widespread [69–71]. It is worthily employed in combination with the above cited survey methods aimed at clarifying and prioritizing the needs customers are aware of. QFD is used as a method to relate the customer demands to the engineering requirements in the early stage of New Product Development (NPD) in order to maximize the satisfaction of the end user. The task is carried out by introducing quantitative variables and Likert scales to characterize the extent of customer satisfaction, the technical performances and their interplay. The intertwining among the variables is represented through a suitable diagram, namely the House of Quality (HoQ).

Despite its deliberate domain of application, some approaches have been developed in order to use QFD for product planning tasks. In [72] QFD and Design Structure Matrix were used to assist the designers in understanding customer needs and planning the early stage of product conceptualization. A market-driven design system was proposed in [73] to integrate QFD technique with commercial analysis. As claimed in [69], the suggested approach allows to concentrate the design efforts on particular product features, with the intended scope of maximizing the expected market appraisal. More recently, Ulrich and Eppinger [74] illustrated a methodological approach for establishing the relative importance of emerging customer needs. Nevertheless such kind of contributions did not result capable to overcome the limitation of optimization methodologies in investigating a narrow space of product profiles with poorly creative results.

In order to describe with more rigour the determinants of customers perceived value, Kano et al. [75] developed a two-dimensional model that relates the degree of satisfaction provided by each attribute according to the offering level it is supplied. Kano introduced three categories for the customer requirements which effectively play a role within the delivery of satisfaction: must-be, one-dimensional and attractive. The most appropriate class of customer satisfaction is determined as a result of tailored VoC questionnaires. The non-linearity between the performance of the attributes and the ensuing satisfaction Kano model allows to highlight is exploited to suggest which characteristics are worth of investments at the maximum extent [76, 77].

The suggested classification is a powerful tool to perform the analysis of the impact played by product features and thus a suitable model to strengthen QFD optimization strategy [78, 79]. The appropriate exploitation of the qualitative information arising from the Kano model, more specifically the categories of customer satisfaction, does not jeopardize the benefits of QFD in treating quantitative variables. Due to indexes representing individual evaluations, the main problem is conversely related to the management of uncertainty. The extent of uncertainty further increases as additional inputs are introduced, leading consequently to marginally reliable outputs to support decision making.

By addressing such problems, Fung et al. [80] have surveyed QFD models to achieve the understanding of uncertainty introduction and propagation, revealing how the relationships between customer requirements and engineering characteristics play a major role. Further on, their research evaluates the effectiveness of linear programming models with fuzzy coefficients to estimate the functional relationships. The employment of fuzzy set theory represents the most diffused approach in the literature for managing the uncertainties and the dynamics of the inputs in QFD; Kahraman et al. [81] proposed a critical review of these applications, but more recent contributions are present. Experiences dealing with uncertainty carried out by means of fuzzy set theory regard also the Kano model [82], as well as its utilization in combination with QFD [83]. From this point of view, although regarded as effective procedures, such ways of managing uncertainty clash with the difficulties in employing fuzzy sets, due to mathematical complexity.

1.3.2.2 Innovations Based on the Delivery of Products and Matched Services

An effective way to map possible patterns of product development regards the investigation of the possible circumstances or phenomena that can impact the use or the behaviour of the artefact. Such monitoring involves all the possible stages of the product existence, from manufacturing to disposal, according to the concept of Life Cycle Engineering. Comprehensive observations of a rich bundle of factors influencing the behaviour of products and users have led towards the delivery of offers including both goods and matched services. The current tendency depicts the need of the companies to manage a greater extent of features and competing factors regarding what could be originally referable to both products and services. Successful business initiatives have offered customers new packages of value attributes by stressing the unique experience faced during the use of certain products. Diffusely, manufacturing firms have endorsed the strategic importance of delivering complementary services, resulting in dramatic business models rethinking [84].

In this framework, Service Product Engineering (SPE) [85] and Product Service Systems (PSSes) [86] have been developed with the aim of generating additional value for products. These methodologies address a growth strategy based on innovation in mature industries, typically by augmenting the overall value for the customer through increased servicing.

Hara et al. [87] developed the notion of Service Product Engineering (SPE) as a means to provide more value to the customer by offering not only products, but also the related services. In their approach, with a tailored VoC task, the designers collect the information about the customer individuality represented by elements of value which constitute the so called "Persona" model. Subsequently each personality is classified into separate groups that summarize the main traits of the customer inclination and internal state. At the same time the characteristics of different service product alternatives are classified through Kano model, heading to assess the impact they have on the receiver's internal state. Eventually, this allows to evaluate the expected customer satisfaction on the basis of the aptitude of different groups. Although dealing with disparate kinds of attributes, such as product features and advantages due to matched services, the main limitation of

SPE approach reflects the inadequacy of VoC in individuating new sources of value than those already outlined.

In the last decade, great attention has been bestowed to PSSes, which represent a particular class of value proposition, that, eventually, can be jointly designed by the enterprise and its customers. The outcome of PSS application is a mix of tangible products and intangible services, which are developed in a synergic way in order to satisfy customer needs [88]. A particular objective of the methodology is represented by sustainability, which is pursued under the assumption that the combined development of product and services results in the reduction of materials consumption in a lifecycle perspective.

Up to date the literature witnesses however only few examples of complete PSSes tasks, due to the lack of a rigorous theory and application procedure [89]. As observed by Baines et al. [90], several contributions provide marginal developments of conventional design methodologies and lack the evaluation of the pursued achievements in practical applications. It follows that not all PSS experiences have fulfilled the expected goals in terms of sustainability and competitiveness, nor the trend of increasing servicing has resulted successful in Business to Customer (B2C) companies [88].

According to the authors' view, the experiences involving New Product/Service Development represent a valuable starting point for companies in renovating their offer, but cover just a niche of profitable initiatives about business model innovation.

1.3.2.3 Experiences of Innovation with NVP

Strategies aimed at performing radical innovation, as those regarding business models, are deemed to play a steadily increasing role within the development of new products [91]. Chesbrough [92] assesses how the innovation of the business model results a fundamental task for firms success, although difficult to be tackled. Kagermann [93] points out how the difficulties faced by many companies deal with their unawareness about the need to innovate their business model and/or reluctance in rethinking their role in the market.

Thus, despite some criticism about the opportunity for established companies to pursue breakthrough innovation strategies [28], the literature acknowledges the advantages gained by reinventing the business model. In such context two issues seem to provoke the most severe limitations for systematically designing business model innovation projects.

The first concern regards the consistent lack of analysis and understanding about business models and their innovation, as claimed by Teece [14]. According to Magretta [8], the literature concerning business models innovation is rich of market triumphs that highlight successful initiatives or intuitions of industrial leaders. On the contrary, limited research has been conducted with the aim of formalizing the determinants that allow the success of business models.

The second concern is related to the different and sometimes contradictory meanings attributed to the term "business model" in the course of time. The diverging interpretations of the concept have consequently led to the emergence of differentiated measures to pursue innovation initiatives. However, according to Keen and Qareshi [94], a general consensus seems to have been reached, representing a business model as a hypothesis "of how to generate value in a customer-driven marketplace". Thus, although the matter may not be undisputed, the concept of value proposition is definitively strictly related with the tasks involving business model innovation. Such interpretation follows in the footsteps of innovation scholars [68], assessing that fundamental breakthroughs require means to deliver a new set of attributes, rather than substantial technological advances. Indeed, disruptive technologies, that underpin the introduction of a new package of attributes, are deemed to show initially lower performances along some dimensions that are valued by established industry customers.

As a result of this diffused vision, the orientation towards customers of innovation programs should not regard the satisfaction of expressed needs, but the research of original and unspoken dimensions of value capable to boost satisfaction. Further insights about the dynamics followed by New Value Proposition (NVP) tasks should therefore be viable to support enterprises that have to undertake customer-centred innovation programs. Definitively, both business and design research involved with New Product Development (NPD) have witnessed a growing interest towards the generation of superior value and experience for the end users [95].

Some investigations have been carried out in order to link the new value attributes to seeded and yet unrevealed needs. In this context, a theoretical background [96] has been built to relate needs theories with the emergence of new attractive customer requirements. In a similar background, studies have been performed to deepen the perception of functional and emotional features of products and services, as well as their relationships with the human needs [97]. As well, Cagan and Vogel [98] have advanced proposals to accomplish NVP strategies based on the interplay of functional and emotional product features. However, the mentioned models result fundamentally descriptive and lack practical indications for the development of products and businesses capable to supply an enhanced customer value.

Within NVP approaches, a branch of business literature e.g. [99] acknowledges the benefits delivered by Blue Ocean Strategy (BOS), fine-tuned by Kim and Mauborgne [100]. Its underpinning theory combines several of the most acknowledged, and previously underlined, concepts within the field of business model innovation: proposition of unprecedented value, redefinition of market boundaries, transition from current industrial standards, etc.

The main assumption of the BOS is that, as supported by the success of mixed Product/Service development initiatives, all traditional industries are already very competitive and capable to oversupply the current demand, needing to look elsewhere for business opportunities. Seeking to 'beat the competition' typically leads to ever-finer segmentation and specialization, price pressure and negative effects on margins. This strategy intends to bring towards the definition of product

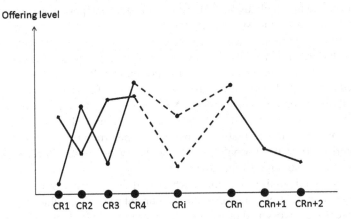

Fig. 1.4 The value curve as a model to represent the dimensions on which alternative profiles compete

characteristics that determine an unprecedented value curve, which strongly differs from the one representing the industrial standard.

The NVP tools that are introduced within the BOS include fundamentally the strategy canvas, graphically depicted through the value curve, and the Four Actions Framework, schematized through the Eliminate Reduce Raise Create (ERRC) Grid. The strategy canvas consists in the general ideas for developing a novel product profile (strategic "move" in BOS jargon). Besides, the value curves stand for the graphical representation of the relative performances of products or services across the relevant factors of competition. The diagram, as that presented in Fig. 1.4, provides a clear vision of the various dimensions of competition (schematized in terms of Customer Requirements—CRs) pertaining two or more alternative product profiles (depicted in grey and black in the graph).

A new curve is built by proper modifications of the current product/service attribute performances and by the introduction of previously ignored properties, throughout the employment of the Four Actions Framework. More in detail, Kim and Mauborgne remark how the Four Actions are applied to product attributes contributing to the buyer's perceived value:

- the eliminate action concerns factors the pertinent industry has long competed on and that do not represent anymore a source of competitive advantage in terms of customer value;
- the reduce action is related to product/service attributes that are overdesigned and that could be provided at much lower performance without affecting perceived value;
- the raise action consists in increasing the performance of certain attributes well above the current industry standard, breaking the compromise with other features of the value curve;
- the create action aims at introducing brand new sources of value for customers.

However, against acknowledged ideas and quite supported assumptions, the BOS currently lacks the systematic paths to envisage innovative products and services, since the introduced tools are elegant to describe past successes, but they are not really prescriptive [101, 102], i.e. they provide just vague indications about the space where to look for new market opportunities. Kim and Mauborgne have illustrated a rich set of case studies from a wide range of industrial sectors, in order to show the strengths of their strategy. Since the authors do not explain how the method has been developed [103], and specifically how they selected the case studies, it has been argued that is not possible to determine whether the examples have contributed to the formulation of the theory or if they have been chosen because they fit the logic of the strategy.

From the applicability viewpoint, whereas it is relatively simple, by benchmarking the competition, to investigate the current relevant product features to be properly removed, worsened or enhanced, the proposition of new valuable product attributes represents a severe challenge [104]. Indeed, it has been argued that the strategy offers just useful visual tools to represent the ideas for exploiting business opportunities, whilst it misses proper guidelines in order to select successful value propositions among multiple alternatives [105]. As a consequence, assessing a strategy canvas results in a difficult matter [106, 107]. Several scholars [108–110] have attempted to improve the robustness of the process of building the strategy canvas, taking into account the extent of importance levels attributed to competition factors in terms of customer perceived value. However, these measures can be adopted just after the relevant business features have been identified and defined, so when the range of possible choices has already been consistently reduced and the actions to be applied have just to be prioritized.

Further matters about BOS applicability and reliability concern the:

- choice and the correct definition of competing factors to be subjected to the actions [111];
- limited rigor about the intended purposes, e.g. regarding the aim of achieving both differentiation and low cost and the exposition of case studies whereas prices have steadily grown;
- the coherence about the strategy recommendations and the illustrated examples, e.g. the claimed need to use all the four actions and strategy canvas showing just Raise and Reduce measures.

Additionally the authors claim that numerous NVP cases, resulting in clamorous flops, can be explained ex-post as outcomes of the application of BOS tools. Thus, the techniques and the basic ideas of Kim and Mauborgne's diffused consulting strategy do not result sufficient to carry out NVP initiatives leading to roaring successes.

The achievement of a robust strategy to support NVP tasks cannot therefore disregard a more careful appraisal of the dynamics followed by successful marketed items and wrong business ideas, since, as remarked by Boztepe [112], the individuation of the proper user factors to be considered in order to provide greater value still remains an open issue.

1.3.3 Summary of the Open Issues Within Product and Process Reengineering

Given the above presented survey of reengineering methodologies involving both product/service and processes, we can state that:

- different kinds of innovation happen according to different stages of product maturity and market penetration;
- the focus on customer value represents a relevant issue within industrial redesign activities;
- traditional BPR initiatives, with the unique aim of reducing process expenditures, show relevant risks due to the disregard of customers, stakeholders and employees;
- beyond the great amount of aspects to be taken into account within reengineering experiences, qualitative variables and uncertainty matters constitute a considerable hurdle in reliably supporting decision making by the means of mathematical models;
- the efforts paid to gather information about customer preferences often result in marginal innovation obtained throughout optimization means;
- within NPD, the proposition of original product profiles characterized by new packages of value attributes, meets the need, expressed by management science community, to design business model innovations strongly differentiating from current market offer;
- among successful NVP experiences, a particular branch involves the integration of servicing within the offer of manufactured products; however no procedure has been acknowledged as a suitable practice for achieving such kind of value innovations;
- generally speaking, the patterns followed along the definition of appealing product profiles lack a comprehensive understanding, which is required in order to formalize methodologies for the generation of superior customer value.

1.4 Purpose of the Book

In this context the development of IPPR reflects the need to plug the gap of the theory about redesign of products and processes and represents a contribution for industries in terms of a support mechanism for undertaking conscious decisions about the directions to be followed within innovation tasks.

As summarized in Fig. 1.5, IPPR orientates the choices to be made along reengineering activities, on the basis of value criteria, by answering the following questions:

- what should be changed in the business process, built as a sequence of industrial phases, in order to effectively fulfill customer requirements or even boost the perceived satisfaction?
- what should be changed in the mix of offered products and services, depicted as a set of attributes, in order to deliver superior value?

Fig. 1.5 Models of the problems to be solved in a BPR initiative based on value innovation

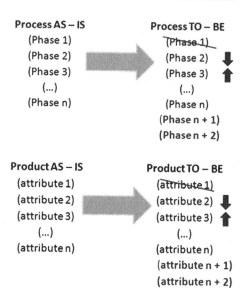

* which investments and business modifications should be prioritized for implementing the new design of the product or the process?
* which established methodologies and tools can result appropriate for the business transformation, according to the new ideas?

In order to identify the most favourable patterns to rethink the process and the products, thus guiding to the generation of the feasible technical and organizational solutions, it is required to carry out an insightful analysis of the current business by performing:

* the investigation of the AS-IS business process at both the economical and technical levels;
* the survey of currently met customer needs and expectations and the individuation of further requirements to be fulfilled with reference to product lifecycle;
* the identification of the bottlenecks in the generation of the value, determining if the product or the process are liable of the main business difficulties.

From a methodological point of view, the previous tasks require the definition of suitable techniques and tools, such as:

* modeling techniques capable to summarize the whole set of information and data pertaining different domains;
* value assessment metrics by which to perform the analysis of the business process focusing on the value delivered to the customer;
* criteria to guide NVP tasks for products and services;
* instruments to support the technical implementation of the reengineering initiatives.

References

1. Davenport TH (1993) Process innovation: reengineering work through information technology. Harvard Business School Press, Boston
2. Hammer M, Champy J (1993) Reengineering the corporation: a manifesto for business revolution. Harper Collins, New York
3. Johansson HJ (1993) Business process reengineering: breakpoint strategies for market dominance. Wiley, New York
4. Kim J, Wilemon D (2002) Sources and assessment of complexity in NPD projects. Res Dev Manag 33(1):16–30
5. Koen P, Ajamian G, Burkart R, Clamen A, Davidson J, D'Amore R, Elkins C, Herald K, Incorvia M, Johnson A, Karol R, Seibert R, Slavejkov A, Wagner K (2001) Providing clarity and a common language to the 'fuzzy front end'. Res Technol Manag 44(2):46–55
6. Smith PG, Reinertsen DG (1998) Developing products in half the time, 2nd edn. Wiley, New York
7. Mahajan V, Srinivasan R, Wind J (2002) The dot.com retail failures of 2000: were there any winners? J Acad Mark Sci 30(4):474–486
8. Magretta J (2002) Why business models matter. Harv Bus Rev 80:86–92
9. Amit R, Zott C (2001) Value creation in e-business. Strateg Manag J 22(6/7):493–520
10. Chesbrough H, Rosenbloom RS (2002) The role of the business model in capturing value from innovation: evidence from xerox corporation's technology spin-off companies. Ind Corp Change 11(3):529–555
11. Francis D, Bessant J (2005) Targeting innovation and implications for capability development. Technovation 25(3):171–183
12. Johnson MW, Christensen CM, Kagermann H (2008) Reinventing your business model. Harv Bus Rev 12(12):57–68
13. Kuratko DF, Audretsch DB (2009) Strategic entrepreneurship: exploring different perspectives of an emerging concept. Entrepreneurship Theory Pract 33:1–17
14. Teece DJ (2010) Business models, business strategy and innovation. Long Range Plan 43(2/3):172–194
15. Markides C (2006) Corporate refocusing. Bus Strategy Rev 4(1):1–15
16. Gotzsch J, Channaron JJ, Birchall D (2006) Product development with a focus on attractive product expression: an analysis of case studies. Int J Prod Dev 3(3/4):1–17
17. Kim WC, Mauborgne R (2004) Blue ocean strategy. Harv Bus Rev 8(10):69–80
18. Barnes C, Blake H, Pinder D (2009) Creating and delivering your value proposition: managing customer experience for profit, 1st edn. Kogan Page, London
19. Miles LD (1949) How to cut costs with value analysis. McGraw Hill Publishing, New York
20. Tontini G (2003) Deployment of customer needs in the QFD using a modified Kano model. J Acad Bus Econ 2(1):103–113
21. Abernathy WJ, Utterback JM (1978) Patterns of innovation in industry. Technol Rev 80(7):40–47
22. Pahl G, Beitz W (2007) Engineering design: a systematic approach, 3rd edn. Springer, London
23. Lacroix RN (2006) Business plan for a modern and efficient business. In: 5th entrepreneurship workshop at Harokopio University, Athens, 5–6 December 2006
24. Brockman K (2008) How to perform a feasibility study and market analysis to determine if an ancillary service makes sense. Orthop Clin N Am 39(1):5–9
25. Mačikėnas E, Makūnaitė R (2008) Applying agent in business evaluation systems. Inf Technol Control 37(2):101–105
26. Kim WC, Mauborgne R (2001) Knowing a winning business idea when you see one. Harv Bus Rev 78(5):129–138
27. Kusiak A, Tang CY (2006) Innovation in a requirement life-cycle framework. In: 5th international symposium on intelligent manufacturing systems, Sakarya, 29–31 May 2006

28. Barwise P, Meehan S (2006) In the box innovation'. Bus Strategy Rev 17(2):68–73
29. Foster NR (1986) Innovation: the attacker's advantage. Summit Books, New York
30. Christensen CM (1992) Exploring the limits of the technology S-curve. Part I: component technologies. Prod Oper Manag 1(4):334–357
31. Christensen CM (1992) Exploring the limits of the technology S-curve. Part II: architectural technologies. Prod Oper Manag 1(4):358–366
32. Holland D, Kumar S (1995) Getting past the obstacles to successful reengineering. Bus Horizons 38(3):79–85
33. Ozcelik Y (2010) Do business process reengineering projects payoff? Evidence from the United States. Int J Proj Manag 28(1):7–13
34. Chen C, Yan W (2008) An in-process customer utility prediction system for product conceptualization. Expert Syst Appl 34(4):2555–2567
35. McGrath ME, Anthony MT, Shapiro AR (1992) Product development success through product and life-time excellence. Butterworth, Boston
36. Khosrowpour M (2006) Cases on information technology and business process reengineering. Idea Group Publishing, Hershey
37. Holweg M (2007) The genealogy of lean production. J Oper Manag 25(2):420–437
38. Herron C, Braiden PM (2006) A methodology for developing sustainable quantifiable productivity improvement in manufacturing companies. Int J Prod Econ 104(1):143–153
39. Woodruff RB (1997) Customer value: the next source for competitive advantage. J Acad Mark Sci 25(2):139–153
40. Knights D, Wilmott H (2000) The reengineering revolution?: critical studies of corporate change. SAGE Publications, London
41. Hall G, Rosenthal J, Wade J (1993) How to make reengineering really work. Harv Bus Rev 71(6):119–131
42. Terziovski M, Fitzpatrick P, O'Neill P (2003) Successful predictors of business process reengineering (BPR) in financial services. Int J Prod Econ 84(11):35–50
43. Edwards C, Peppard J (1994) Forging a link between business strategy and business reengineering. Eur Manag J 12(4):407–416
44. Ibusuki U, Kaminski PC (2007) Product development process with focus on value engineering and target-costing: a case study in an automotive company. Int J Prod Econ 105(2):459–474
45. Chan KK, Spedding TA (2003) An integrated multidimensional process improvement methodology for manufacturing systems. Comput Ind Eng 44(4):673–693
46. Reijers HA, Liman Mansar S (2005) Best practices in business process redesign: an overview and qualitative evaluation of successful redesign heuristics. Omega 33(4):283–306
47. Kaplan RS, Norton DP (1996) Using the balanced scorecard as a strategic management system. Harv Bus Rev 74(1):75–85
48. Voelpel SC, Leibold M, Eckhoff RA, Davenport TH (2006) The tyranny of the balanced scorecard in the innovation economy. J Intellect Cap 7(1):43–60
49. Remenyi D, Heafield A (1996) Business process reengineering: some aspects of how to evaluate and manage the risk exposure. Int J Proj Manag 14(6):349–357
50. Chan SL, Choi CF (1997) A conceptual and analytical framework for business process reengineering. Int J Prod Econ 50(2–3):211–223
51. Irani Z, Hlupic V, Baldwin LP, Love PED (2000) Reengineering manufacturing processes through simulation modelling. Logist Inf Manag 13(1):7–13
52. Lambert JH, Jennings RK, Joshi NN (2006) Integration of risk identification with business process models. Syst Eng 9(3):187–198
53. Min DM, Kim JR, Kim WC, Min D, Ku S (1996) IBRS: intelligent bank reengineering system. Decis Support Syst 18(1):97–105
54. Williams A, Davidson J, Waterworth S, Partington R (2003) Total quality management versus business process reengineering: a question of degree. Proc Inst Mech Eng Part B J Eng Manuf 217(B1):1–10

55. Wang KJ, Lin YS (2007) Resource allocation by genetic algorithm with fuzzy inference. Expert Syst Appl 33(4):1025–1035

56. Mahdavi I, Shirazi B, Solimanpur M (2010) Development of a simulation-based decision support system for controlling stochastic flexible job shop manufacturing systems. Simul Model Pract Theory 18(6):768–786

57. Völkner P, Werners B (2000) A decision support system for business process planning. Eur J Oper Res 125(3):633–647

58. Gregoriades A, Sutcliffe A (2008) A socio-technical approach to business process simulation. Decis Support Syst 45(4):1017–1030

59. Llamas-Alonso MR, Jiménez-Zarco AI, Martínez-Ruiz MP, Dawson J (2009) Designing a predictive performance measurement and control system to maximize customer relationship management success. J Mark Channels 16(1):1–41

60. Azadeh A, Nassiri S, Asadzadeh M (2010) Modeling and optimization of a purchasing system in uncertain environments by an integrated fuzzy business process simulation and data envelopment analysis: a novel approach. In: Proceedings of the spring simulation multi conference (SpringSim'10), pp 1–8, Orlando, 11–15 April 2010

61. He H, Jiang L, Li B (2009) Business process reengineering risk assessment based on a new improved FAHP. In: Proceedings of the Asia-Pacific conference on information processing, vol 2, pp 278–281, Shenzhen, 18–19 July 2009

62. Ramirez R, Melville N, Lawler E (2010) Information technology infrastructure, organizational process redesign, and business value: an empirical analysis. Decis Support Syst 49(4):417–429

63. Grover V (1999) From business reengineering to business process change management: a longitudinal study of trends and practices. IEEE Trans Eng Manag 46(1):36–46

64. Grint K (1994) Reengineering history, social resonances and business process reengineering. Organization 1(1):179–201

65. Charles W, Zamzow Jr (2008) Business process-reengineering: 7 critical success factors for a smooth transformation of your organization processes. Wordclay, Bloomington

66. Van Kleef E, Van Trijp HCM, Luning P (2005) Consumer research in the early stages of new product development: a critical review of methods and techniques. Food Qual Preference 16(3):181–201

67. Ulwick AW (2002) Turn customer input into innovation. Harv Bus Rev 80(1):91–97

68. Bower JL, Christensen CM (1995) Disruptive technologies: catching the wave. Harv Bus Rev 73(1):43–53

69. Yan W, Khoo LP, Chen C (2005) A QFD-enabled product conceptualisation approach via design knowledge hierarchy and RCE neural network. Knowl Based Syst 18(6):279–293

70. Day RG (1993) Quality function deployment—linking a company with its customers. ASQC Quality Press, Milwaukee

71. Hauser JR, Clausing D (1988) The house of quality. Harv Bus Rev 66(3/4):63–73

72. Houvila P, Cere'n KJ (1998) Customer-oriented design methods for construction projects. J Eng Des 9(3):225–238

73. Harding JA, Popplewell K, Fung RYK, Omar AR (2001) An intelligent information framework relating customer requirements and product characteristics. Comput Ind 44(1):51–65

74. Ulrich K, Eppinger SD (2004) Product design and development, 3rd edn. McGraw-Hill, New York

75. Kano KH, Hinterhuber HH, Bailon F, Sauerwein E (1984) How to delight your customers. J Prod Brand Manag 5(2):6–17

76. Conklin M, Powaga K, Lipovetsky S (2004) Customer satisfaction analysis: identification of key drivers. Eur J Oper Res 154(3):819–827

77. Chen C, Chuang M (2008) Integrating the Kano model into a robust design approach to enhance customer satisfaction with product design. Int J Prod Econ 114(2):667–681

78. Matzler K, Hinterhuber HH (1998) How to make product development projects more successful by integrating Kano's model of customer satisfaction into quality function deployment. Technovation 18(1):25–38

79. Tan KC, Shen XX (2000) Integrating Kano's model in the planning matrix of quality function deployment. Total Qual Manag 11(8):1141–1151

80. Fung RYK, Chen Y, Tang J (2006) Estimating the functional relationships for quality function deployment under uncertainties. Fuzzy Sets Syst 157(1):98–120

81. Kahraman C, Ertay T, Büyüközkan G (2006) A fuzzy optimization model for QFD planning process using analytic network approach. Eur J Oper Res 171(2):390–411

82. Lee YC, Huang SY (2009) A new fuzzy concept approach for Kano's model. Expert Syst Appl 36(3):4479–4484

83. Chen LH, Co WC (2008) A fuzzy nonlinear model for quality function deployment considering Kano's concept. Math Comput Model 48(3–4):581–593

84. Aurich JC, Mannweiler C, Schweitzer E (2010) How to design and offer services successfully. CIRP J Manuf Sci Technol 2(3):136–143

85. Kimita K, Shimomura Y, Arai T (2009) A customer value model for sustainable service design. CIRP J Manuf Sci Technol 1:254–261

86. Mont OK (2002) Clarifying the concept of product-service system. J Clean Prod 10(3):237–245

87. Hara T, Arai T, Shimomura Y, Sakao T (2009) Service CAD system to integrate product and human activity for total value. CIRP J Manuf Sci Technol 1:262–271

88. Tukker A, Tischner U (2006) Product-services as a research field: past, present and future. Reflections from a decade of research. J Clean Prod 14(17):1552–1556

89. Lasalle D, Britton TA (2003) Priceless: turning ordinary products into extraordinary experiences. Harvard Business School Press, Boston

90. Baines TS, Braganza A, Kingston J, Lockett H, Martinez V, Michele P, Tranfield D, Walton I, Wilson H (2007) State-of-the-art in product-service systems. Proc Inst Mech Eng Part B J Eng Manuf 221(10):1543–1552

91. Sandstrom C, Bjork J (2010) Idea management systems for a changing innovation landscape'. Int J Prod Dev 11(3/4):310–324

92. Chesbrough H (2010) Business model innovation: opportunities and barriers. Long Range Plan 43(2/3):354–363

93. Kagermann H (2008) Reinventing your business model. Harv Bus Rev 86(6):1–11

94. Keen P, Qureshi S (2006) Organizational transformation through business models: a framework for business model design. In: Proceedings of the 39th Hawaii international conference on system sciences, Kōloa, 4–7 January, 2006

95. Søndergaard HA (2005) Market-oriented new product development: how can a means-end chain approach affect the process? Eur J Innov Manag 8(1):79–90

96. Ward D, Lasen M (2009) An overview of needs theories behind consumerism. J Appl Econ Sci 4(1):137–155

97. Crilly N, Moultrie J, Clarkson PJ (2004) Seeing things: consumer response to the visual domain in product design. Des Stud 25(6):547–577

98. Cagan J, Vogel CM (2002) Creating breakthrough products: innovation from product planning to program approval. Prentice-Hall, Upper Saddle River

99. Leavy B (2005) Value pioneering—how to discover your own "blue ocean": interview with W. Chan Kim and Renée Mauborgne. Strategy Leadersh 33(6):13–20

100. Kim WC, Mauborgne R (2005) Blue ocean strategy: how to create uncontested market space and make competition irrelevant. Harvard Business School Press, Boston

101. Aspara J, Hietanen J, Parvinen P, Tikkanen H (2008) An exploratory empirical verification of blue ocean strategies: findings from sales strategy. In: Proceedings of the eighth international business research (IBR) conference, Dubai, 27–28 March, 2008

102. Parvinen P, Aspara J, Kajalo S, Hietanen J (2010) An exploratory empirical examination of blue ocean practices in sales management'. Paper presented at the 26th industrial marketing and purchasing group (IMP) conference, Budapest, 1–3 September, 2010

103. Khalifa S (2009) Drawing on students' evaluation to draw a strategy canvas for a business school. Int J Educ Manag 23(6):467–483
104. Sheehan NT, Vaidyanathan G (2009) Using a value creation compass to discover blue oceans. Strategy Leadersh 37(2):13–20
105. Raith MG, Staak T, Wilker HM (2007) A decision-analytic approach to blue-ocean strategy development. In: Operations research proceedings 2007, Saarbrücken, 5–7 September, 2007
106. Abraham S (2006) Blue oceans, temporary monopolies, and lessons from practice'. Strategy Leadersh 34(5):52–57
107. Jackson SE (2007) Design, meanings and radical innovation: a meta-model and a research agenda'. J Prod Innov Manag 25(5):436–456
108. Kim C, Yang KH, Kim J (2008) A strategy for third-party logistics systems: a case analysis using the blue ocean strategy. Omega 36(4):522–534
109. Narasimhalu AD (2007) Designing the value curve for your next innovation'. In: Proceedings of PICMET 2007 conference, Portland, 5–9 August, 2007
110. Ziesak J (2009) Wii innovate—how Nintendo created a new market through the strategic innovation Wii. GRIN Verlag, Munich
111. Hong ANH, Chai DLH, Ismail W (2011) Blue ocean strategy: a preliminary literature review and research questions arising. Aust J Basic Appl Sci 5(7):86–91
112. Boztepe S (2007) Toward a framework of product development for global markets: a user-value-based approach. Des Stud 28(5):513–533

Chapter 2
IPPR Methodological Foundations

2.1 Introduction

The analysis and solution model underpinning IPPR consists in a set of activities that are organized along a common path for each class of reengineering problems, as clustered in the previous Chapter.

Basically IPPR method follows a well established logic which is universally acknowledged as a standard to analyze and solve technical problems. It is grounded on three main phases: (i) situation analysis and representation of the relevant information; (ii) identification of the system criticalities; (iii) individuation of the suitable solving directions. The aim of IPPR is to perform the step-by-step procedure with a constant orientation towards what concerns customer value and perceived satisfaction. To this end, the whole body of the methodology suggests suitable tools and techniques.

However, given the consolidated logic adopted by IPPR (i.e. *analysis* of the problem, *diagnosis* of the reengineering opportunities, *synthesis* of the solutions), each task can be performed by the usage of alternative instruments dedicated to the design of products and processes. Thus, the reader can use his/her own body of knowledge to carry out the activities consistent with IPPR, with regards to his/her competencies in the fields of business process reengineering and new product development. Otherwise, the user can benefit from the original tools illustrated in Chap. 3, which highlights the preferred employment of value-oriented instruments for each step of the methodology.

According to the objectives of this Chapter, the introductory parts of the Subsections belonging to 2.2 report an overall description of IPPR by providing an overview of the main methodological steps and their partial outputs. Subsequently, the remaining content of the Subsections reports a detailed description of the tasks, activities, expected results foreseen by each step of the methodology. As a whole, the presentation of the coverage is organized on the basis of the classification of the business problems already introduced in Chap. 1.

F. Rotini et al., *Re-engineering of Products and Processes*,
Springer Series in Advanced Manufacturing, DOI: 10.1007/978-1-4471-4017-7_2,
© Springer-Verlag London 2012

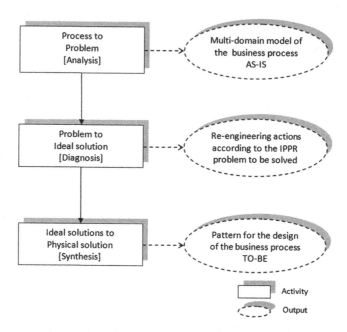

Fig. 2.1 Workflow and partial outputs of IPPR methodology

As a result of the description of IPPR structure, the main activities to be carried out are summarized in Sect. 2.3.

2.2 The Logic and the Structure of IPPR: Steps, Activities and Outcomes

IPPR methodology leads the user to the identification of feasible process/product innovations by means of an analysis and solution path based on three main steps. The workflow of activities and the arising outputs are depicted in Fig. 2.1.

However, in order to successfully carry out the depicted activities, IPPR practitioners are requested to preliminarily acquire the information about the problem to be investigated. With this aim, the Sect. 2.2.1 discusses the objectives to be attained with regards to the collection of the essential elements of knowledge.

The aim of the *Process to problem* step is to obtain an exhaustive description of the AS-IS situation by investigating the industrial operations and their outputs. The result of this phase is constituted by a model of the business process capable to represent all the aspects related to both the functional and economic domains. Such a multidimensional approach allows to manage the cross-disciplinary nature of the

Table 2.1 Organization of the content related to each phase of IPPR methodology within the present Chapter in function of the class of business process problem to be addressed

	Problem to process	Process to ideal solution	Ideal solution to physical solution
Class 1 and Class 2	2.2.2.1 2.2.2.2 2.2.2.3	2.2.3.1	2.2.4.1
Class 3	2.2.2.2 2.2.2.3	2.2.3.2	2.2.4.2

business process. This is the key feature enabling a comprehensive analysis of a large amount of common industrial problems.

The loss of competitiveness of a business process occurs when the provided outputs are no longer able to satisfy the customer expectations, nor to attract market segments through appealing and original product designs. The causes that determine such situation have been already extensively described in Chap. 1 and, generally speaking, they may be related to aspects falling into the sphere of industrial process and/or of the delivered product. Such causes represent what we can call *value bottlenecks*, since they somehow impact (negatively) the customer perceived value.

The second step, named *Problem to Ideal solution*, is focused on the clear identification of the recalled value bottlenecks and eventually of potential innovation opportunities. Moreover, once the critical aspects of the business process have been analyzed, proper reengineering actions are defined in order to remove the value bottlenecks and preserve or regain the market competitiveness. These guidelines are expressed in the form of new process requirements for the problems belonging to the class 1 and 2, while they are depicted as directions for the transformation of product profiles, with reference to the class of problem 3. The emerging hints represent the inputs of the subsequent design activities which are aimed at identifying suitable technical solutions for the implementation of the ideas of the new process or product.

The last step, namely *Ideal solution to Physical solution*, suggests the suitable and acknowledged instruments to support the design activities of the physical solutions concerning the introduction of new industrial process phases, the improvement of the existing ones, the reorganization of the resources allocation programs, the production of innovative items and the delivery of novel services.

The sequences of activities summarized in Fig. 2.1 are customized according to the business process problem that should be addressed. Table 2.1 indicates the sections of the present Chapter in which the reader can find the relevant criteria to shape each step of the IPPR methodology according to the class of reengineering problems defined in Chap. 1.

Table 2.2 Checklist providing the overall set of relevant information to be gathered according to the class of problem to be faced

Problem to be solved	Information to be gathered
Class 1 and Class 2	Phases of the business process
	Flows of materials, energy and information
	Elapsed duration of each phase, labour time, dead times
	Involved technologies
	Occupied space
	Involved human skills and knowledge
	Other phase expenditures
	Control and evaluation parameters governing each phase
	Customer requirements and their relevance in determining the customer perceived value
	Contribution of each phase in determining the product requirements
Class 2	Determinants for delighting the customer
	Determinants for avoiding the customer dissatisfaction
Class 3	Product attributes of the treated product and of the competing ones
	Performances levels at which the product attributes are delivered
	Kind of benefits perceived by the user in delivering product attributes

2.2.1 Performing Information Gathering for IPPR

The information gathering is a preliminary activity to be performed in order to widen the knowledge of the IPPR user about the business problem to be treated. The additional information to be collected with the aim of carrying out the subsequent tasks in a more rigorous way strongly depends on the nature and the role of the practitioner (product manager, analyst, CEO, researcher, etc.) and thus on the main individual lacks of knowledge.

In order to address the sources of information to be preliminarily consulted, Table 2.2 summarizes the aspects to be treated within IPPR with reference to process and product reengineering.

Commonly, the activity is carried out by taking in consideration several information sources. At the beginning of the information acquisition, sources like books, reports, manuals and catalogues play a significant role for the definition of the background of the industrial sector to be analyzed [1]. Subsequently, more detailed and explicit information can be extracted through the consultation of domain experts and involved personnel [2].

The ideal result of the information acquisition would be the extraction and codification of tacit knowledge, which plays a significant role especially within the description of processes, by highlighting human practices when performing operations. The concept of tacit knowledge was introduced by Polanyi [3], who defined it as personal, with no possibility to be codified. Since then, the possibility of acquiring and disseminating tacit knowledge is a very debated issue. Many scholars, such as Nonaka [4], have developed Polanyi's conception of tacit knowledge in a practical direction to enhance organizational knowledge creation, assessing the possibility to elicit it. Coherently to this vision and purpose, the task of acquiring tacit knowledge implies to meet directly the employees; the consultation on the shop floor recalls the concept of "gemba", a Japanese term meaning "the place where the truth can be found", firstly introduced by Mazur [5] within Quality Function Deployment (QFD) [6].

Also when the attempt of collecting elements of tacit knowledge results an excessively challenging and time-consuming task, it is recommended to take into account the viewpoint of multiple experts. However this can result in contradicting issues arising from overlapping competencies of the involved specialists. In order to overcome the difficulties dictated by the emergence of conflicting visions, different approaches can be chosen:

- a final conjoint consultation of the experts can be organized to conciliate the diverging viewpoints;
- IPPR steps 1 and 2 can be performed separately by multiple experts and then the resulting reengineering directions are compared and integrated;
- with reference to the classes of problem 1 and 2, which employ more quantitative coefficients, statistical tools generally dedicated to deal with uncertainty can be favorably employed to the outcomes of steps 1 and 2, leading to a "best" description and analysis of the process.

2.2.2 Process to Problem Phase

The aim of the first step is to schematize the business process into a general model of the problem, allowing to perform the subsequent analysis steps foreseen by the IPPR methodology.

Such a model describes how the system works in both the technical and economic domain. It summarizes the sequence of the performed phases and their mutual relationships expressed through the flows of inputs/outputs and involved resources such as: material, energy and information, technologies, human skills and know-how, elapsed times and monetary expenditures.

The final outputs of the process are represented by the customer requirements which are fulfilled by the manufactured products and delivered services. These attributes are depicted throughout their performance or offering level and their relevance in determining customer satisfaction and/or avoiding buyer's discontentment. The Fig. 2.2 shows, in a schematic way, the input data and a conceptual representation of the output model provided by this step.

In order to accomplish the above mentioned objectives, the *Process to Problem* step requires the execution of the following specific activities:

- industrial process modeling;
- product information elicitation;
- product modeling.

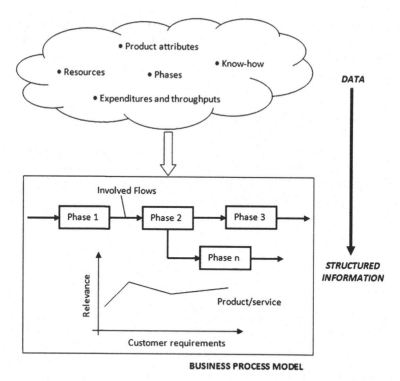

Fig. 2.2 The "Process to problem" step brings to the definition of a model of the business process (from both the industry and the customer perspective) in both the technical and economical domains. This model is used to perform all the subsequent analysis tasks

Here in the followings these tasks are described in detail, according to the class of reengineering problems to be addressed.

2.2.2.1 Process Modeling

For the problems belonging to the classes 1 and 2, the collected data have to be organized in order to build the process model, a structured representation of the AS-IS situation. Several representation methods with diverging formalisms are available in the literature to support the modeling of industrial activities. However, the various techniques significantly differ in the ability to model the system according to different domains and perspectives. Some techniques focus primarily on the data flows, others on the deployed functions or on the assigned roles of human resources within the process, etc. [7, 8].

A customized multi-domain model, presented in Chap. 3, is suggested to represent the information and data needed to implement the IPPR methodology. Its advantages arise as a result of the hybridization of different modeling techniques, each one tailored to represent different facets of a business process.

Whereas the user would prefer the employment of mastered modeling techniques, the representation of the process has to include at least the following important aspects:

- *Functions*: the model has to report the process phases in terms of performed functions, input and output flows;
- *Multi-domain features*: for each phase of the industrial process, the model has to summarize the involved flows of resources in both the technical (i.e. flows of energy, materials and information) and economical (i.e. monetary flows or equivalent indicators) domains;
- *Control variables and performances*: the model has to allow a clear representation of the control parameters governing each process phase (e.g. cutting speed of a machine tool, bill of materials), as well as the required performances.

2.2.2.2 Product Information Elicitation

The information that is schematized within the process model (classes 1 and 2) supports the identification of a large set of features that the product should have. Indeed, the designed transformations of channeled resources into desired outcomes are justified in terms of the fulfillment of the customer requirements. However, in order to represent a comprehensive record of the process outputs in terms of the elements that currently participate to the building of customer value, suitable checklists are proposed within this step of IPPR.

Furthermore, with the objective of accurately characterizing the business process, the elicitation is a crucial activity of the relationships existing among the phases and the terms contributing to the perceived customer value. At the firm level, the phases can be considered like the segments that constitute the value chain, as defined in the literature by Porter [9] and some other scholars. According to this concept, each function performed along the investigated process contributes in fulfilling the characteristics of the final product or service, thus in the generation of value. Basically, the extent of such a contribution depends on the number of properties of the elaborated inputs that are modified by the function, as well as by the magnitude of such changes. In the context of product development strategies, the recalled QFD investigates the interplay among customer expectations and engineering characteristics that meet the needs of the end-user. With a similar logic, the proposed task requires mapping the features underlying the accomplishment of each customer requirement. Subsequently the phases, properly identified in the modeling step, that modify or deal with those features are monitored by the business process experts in order to define their accounted ratios in fulfilling the customer requirements (CRs). For instance, the requested speed of a courier service is achieved by the correct functioning of all the phases impacting the delivery time of some goods, thus all the operations concerning the scheduling, the warehousing and the transportation of the sent items. The relative contributions

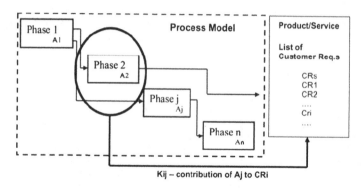

Fig. 2.3 The coefficients k_{ij} represent the contribution of the j-th phase to the satisfaction of the i-th customer requirement

addressed to the j-th phase in ensuring the achievement of the i-th customer requirement (CR) will be further on indicated with the variable k_{ij}. As represented in Fig. 2.3, the coefficients k_{ij} can be evaluated as a correlation between the properties of the objects modified by each function and the CRs of the final product.

With reference to the problems concerning product reengineering (class 3), the objective of the activity is the elicitation of the information related to the dimension of customer satisfaction. A suitable tool is proposed with the aim of individuating the circumstances potentially guiding to the emergence of sources of value, regardless they have been already exploited or not. The structured search should therefore lead to the individuation of a comprehensive set of offered product attributes and, eventually, if required by the case study, to ease the monitoring of the competition. Additionally, the mapping process may allow the discovery of disregarded performances or characteristics, thus facilitating the task of designing a new product profile.

2.2.2.3 Product/Service Modeling

The product model summarizes the offered value profile according to the competing factors of the market where the business process operates. This activity is aimed at identifying the product attributes delivered to the customer, their relevance and role in determining the customer satisfaction.

Within IPPR, different representations are adopted for the process related problems (classes 1 and 2) and product oriented reengineering tasks (class 3).

The first circumstance requires a description of the process output in the perspective of the company, by shedding light on *how much* the product delivers value or avoids dissatisfaction. A basic activity concerns therefore the classification of the customer requirements through a criterion capable to highlight the extent in impacting the customer contentment. As widely acknowledged in

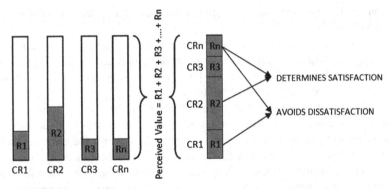

CRi = i-th customer requirement
Ri = extent (relevance) of the i-th CR in determining the customer perceived value

Fig. 2.4 The classification scheme of the customer requirements according to the relevance in determining the customer perceived value and the role played in impacting the customer satisfaction/dissatisfaction

the literature [10], some product characteristics are able to generate satisfaction for the end-user, while the presence of some other characteristics is merely motivated by the need to avoid the customer dissatisfaction. Moreover, the extent in determining the customer satisfaction and/or avoiding dissatisfaction depends on the importance of each product feature within the perceived value. Thus, the twofold properties characterizing the customer requirements suggest the adoption of a scheme (Fig. 2.4) to describe the attributes in terms of:

- the extent (relevance R) at which they impact the customer perceived value;
- the role in determining the customer satisfaction and avoiding the product rejection.

Nevertheless, certain circumstances can invalidate the prerequisites for which the distinction is meaningful between attributes aimed at, respectively, generating satisfaction and avoiding discontentment. Such conditions characterize all business processes intended to fulfill just customer requirements imposed by regulations, standards or requested by the purchaser (i.e. for third-parties or suppliers), as well as merely focused at replicating the performances of the products in the marketplace. Beyond imitation, the last case is common for companies facing the need to achieve certain product characteristics to stay competitive, but whose business is affected by unpredictable external problems (e.g. shortage of materials, soaring prices of certain required resources, etc.). Thus, whenever the reengineering task is oriented towards the achievement of predefined targets, ranging outside the sphere of the company decisions, the qualitative classification of the role played by the product features in impacting the customer satisfaction misses the original sense. According to this assumption, all the business process problems falling into the first class do not require the classification of the customer requirements, being the definition of the relevance scores sufficient to characterize their contribution in building value.

Among the classification hypotheses regarding the different kind of features determining the customer satisfaction, the model employed by IPPR adopts the categories introduced by Kano et al. [10], representing the most established clustering criteria available in the literature.

With the aim of supporting the problems belonging to class 3, a suitable representation of the product profile is proposed, which emphasizes *how* the attributes deliver value and whether their offering level is adequate for the current demand from the customer viewpoint. For the scope of product reengineering, a suitable clustering of the fulfilled attributes supports the identification of the most favorable directions to attain new value profiles. Such a categorization concerns the distinction of the product features according to the functional role played in determining positive outcomes for the customer, in avoiding limitation of undesired effects or in giving rise to the reduction of required resources, with reference to the terms that contribute to "Ideality" as suggested by TRIZ [11].

2.2.3 *Problem to Ideal Solution*

This step is aimed at identifying "what should be changed" in the AS-IS business process in order to increase the benefits for the company, as a result of the enhanced customer value. The customer satisfaction is evaluated as a direct function of the delivered product attributes.

As recalled, according to the classes of reengineering problems defined in Chap. 1, the actions to be undertaken may regard the process, the product or both of them.

The faced difficulties regarding the process may concern the hurdles in entering a new market due to under capacities in providing mandatory product characteristics (class 1) or the loss of competiveness for a consolidated business (class 2). In both circumstances this step is aimed at highlighting the value bottlenecks that hinder the maximization of the customer satisfaction according to the available resources and the buyer demands. This analysis is the starting point for the effective reorganization of the process pursuing the increment of the value delivered to the end-user.

Otherwise, if the lack of competitiveness is due to a product that is definitively no longer capable to appeal the marketplace, it is necessary to define a new value profile. The redesign of the industry outputs can be obtained by rethinking the overall business model and, more specifically, by identifying the product characteristics that can be worthily introduced, emphasized or eventually removed without particular consequences.

With reference to the classification of the industrial problems suggested in the Chap. 1, the implementation of the following tasks is required:

- *Identification of what should be changed in the process*: it is required to solve business problems related to the competitiveness of the industrial process, i.e. problems belonging to the classes 1 and 2.
- *Identification of what should be changed in the product*: it is required to solve problems of product competiveness, i.e. problems falling into the class 3.

In the following paragraphs the activities aimed at identifying what should be changed in the process or in the product, are described in detail.

2.2.3.1 Identification of What Should Be Changed in the Process

Value Engineering, the well known methodology developed by Miles [12], represents a useful starting point with the purpose of identifying the business short-comings. However, according to the considerations performed in Sects. 2.2.2.2 and 2.2.2.3, the value assessment strategy suggested by Miles requires a shift in order to be employed for the aim of IPPR, from the system perspective to the viewpoint of generated customer satisfaction. More precisely, instead of considering the revenues (as a function of the technical performances) provided by the process functions and the spent resources, the generated benefits should be measured in terms of satisfaction for the customer. Within this vision, the logical path followed by IPPR to identify process bottlenecks, is constituted by three main activities as shown in Fig. 2.5. The involved tasks allow the assessment of the phases' worthiness by exploiting the information gathered in the *Process to Problem* step.

With reference to the industrial problems grouped within the class 2, the coefficients k_{ij} give the possibility to evaluate for each phase suitable indexes, namely *Phase Customer Satisfaction* (PCS) and *Phase Customer Dissatisfaction* (PCD), that express the potential to bring customer contentment and the contribution in avoiding dissatisfaction. Such values represent, respectively, the opportunity for a phase to delight customers and the risk to harmfully impact the product perception. Thanks to PCS and PCD it is possible to determine the contribution of each phase to the general customer contentment by means of an indicator named *Phase Overall Satisfaction* (POS), which is assumed as a measure of the benefits provided by the phase.

A review of the literature shows the availability of metrics for the calculation of indexes to evaluate overall appreciation of products as a function of the terms expressing positive and negative evaluations by the customers (such as PCS and PCD, respectively). Commonly the impacts of satisfaction and (avoided) dissatisfaction are related through linear and non-linear equations to the overall satisfaction. Among them, the one receiving the widest consensus has been obtained through a research work performed by Mittal et al. [13] and has been adopted as a reference for the IPPR methodology. The employed equation is non-linear and it states the asymmetric influence of positive and negative evaluations, with a greater role played by dissatisfaction factors in impacting the general customer

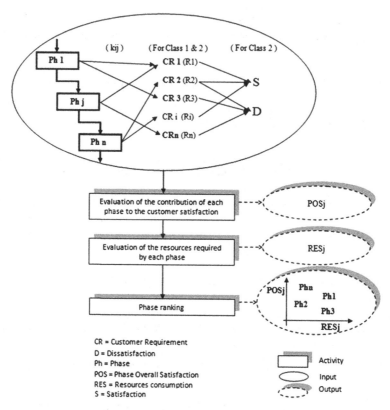

Fig. 2.5 The logical path suggested by IPPR in order to identify what should be changed in a process

contentment. More details about such model and the metrics adopted for the calculation of PCS and PCD coefficients are provided in Chap. 3.

For the business problems categorized within the class 1, due to the missing of diverse contributions to satisfaction and discontentment, the calculation of the POS is performed by taking into account just the k_{ij} coefficients and the relevance indexes R of the attributes.

Once the POS coefficients have been calculated, the next step requires the evaluation of the resources spent by each phase and eventually the estimation of undesired effects resulting as the process is displayed. Within the context of business processes it is suitable to consider the whole range of resources (occupied space, information and know-how, labor, energy, materials, dead times) and measure their extent, in order to use value formulations for calculating quantitative indicators. Long elapsed times to perform the phases represent relevant hurdles for the business, especially for those industries (e.g. fashion), and kind of firms (e.g. third-parties, suppliers) for which timeliness is a crucial competing factor. All the other kinds of employed resources can be compared in terms of the resulting expenditures, so to be evaluated with uniform units of measurement. With regards to significant harmful effects and their consequences (e.g. pollution and measures to limit its impact, noise, need to introduce particular safety systems), they have to be soundly considered as undesired elements within the business process and its phases. In certain cases they can represent even barriers to carry out the process and then to access the market. In other circumstances the harmful functions can occur in the shape of problems affecting the stability of the system, as well as the repeatability of the process.

When monetary costs, meaningful elapsed times and harmful effects coexist in the business process, experts have to weigh their relative relevance, introducing corrective coefficients for the overall estimation of undesired issues. Further on, with the term *resources*, we will indicate the total mix of expenditures, drawbacks and inconveniences that emerge as the phases of the business process are displayed.

Thanks to the results obtained by the previous assessment activities, it is possible to characterize the phases constituting the process in terms of generated benefits versus spent resources. The insightful analysis of the phases leads towards the individuation of the process bottlenecks in a value-wide perspective.

The ratio between POS and the spent resources provides an *Overall Value* (OV) index suitable to globally identify strengths and weaknesses of the process. Those phases showing a high OV can be considered to be tailored to the business process and their employed resources are well spent in generating customer satisfaction, whereas the ones with low scores represent problematic issues.

The conjoint analysis of the POS and the spent resources helps in characterizing the nature of the bottlenecks: when a low OV is due to a high denominator, i.e. great amount of resources, the focus of the reengineering actions must be oriented towards saving policies. Besides, when a poor OV rate is due to a limited contribution to customer satisfaction, the reengineering initiatives should evaluate the opportunity to eliminate the investigated phase by assigning other segments of the process its functions, substitute the technology adopted so far, introduce new features to be fulfilled without a meaningful increase of the needed resources.

With reference to the reengineering problems pertaining class 2, it is possible to perform further evaluations of the phases, by considering separately, with reference to the spent resources, their capability to achieve customer satisfaction and/or to fulfill the basic requirements of the product. A tailored graphical representation introduced within IPPR illustrates the coupled appropriateness of the process phases in delighting customers and avoiding their dissatisfaction.

In the Chap. 3, all the models and formulas to determine the above described parameters are provided, as well as the suggested representation diagrams.

2.2.3.2   Identification of What Should Be Changed in the Product

Such task refers to the most critical activity involved in the New Product Development cycle.

As stated in the previous Chapter, most of the methods developed to support NPD initiatives are based on the so called "Voice of the Customer" (VoC). The business strategy based on this approach entrusts the main choices of innovation task to the end-user of the manufactured product or the delivered service. However, as noticed in Chap. 1, the VoC commonly brings just to the design of incremental innovations, bounded within what customers can already conceive. As a consequence breakthrough solutions, capable to provide substantial competitive advantages, are not diffused.

These evidences have been confirmed also by several scholars in the field of product innovation management. They have demonstrated that business strategies based on the definition of an innovative set of product features for the reference industry of the company, allow to create new market space by performing a New Value Proposition (NVP). Hence, aiming at radically modifying the product, the tools suggested by IPPR for the third class of business problems are oriented towards the achievement of a strategy based on NVP, rather than being addressed to the fulfillment of explicit needs.

With this scope, the most critical aspect related to NVP initiatives is represented by the definition of the new elements of value to be delivered to the customer. As recalled in Chap. 1, the most established approaches, such as those swiveling on increased servicing (PSSs, SPE), represent just a specific strategy within the creation of new value for customers. On the other hand, despite the general appraisal received in the industrial world, the tools proposed by the BOS are affected by scarce applicability, since their nature is predominantly descriptive rather than prescriptive.

In order to overcome the limits of the recalled methodologies an original tool has been developed within IPPR, namely *New Value Proposition Guidelines* (NVPGs). It consists in a set of recommendations capable to orientate the strategic decisions about the definition of a new product profile. The NVPGs, by complementing the general scheme offered by the Four Actions Framework (FAF), identify which value shifts result the most advantageous with respect to the consolidated industrial standards.

The NVPGs have been developed by performing an in-depth analysis of successful market stories, aimed at pointing out common patterns of value evolution. More in detail, the performed survey has individuated which categories of competing factors are preferentially transformed within the treated value transitions, according to the functional features.

On the basis of the performed classification, the NVPGs provide a collection of suggestions in terms of types of new valuable product attributes to create, existing properties to enhance, current features whose performances are viable to be reduced and eventually product characteristics to be eliminated without relevant drawbacks. Hence, the guidelines represent useful recommendations to support value transition tasks within strategies based on business model innovation and NVP.

2.2.4 Ideal Solution to Physical Solution

This step of IPPR addresses the application of the appropriate measures to attain the new process/product specifications, as a result of the *Problem to Ideal Solution* phase. The emerging indications have to be translated in technical objectives and organizational changes, allowing to put in practice all the needed business process modifications.

Thus, the objective of this step is the identification of the proper functions to be performed and the search of appropriate technical solutions for their implementation.

According to the class of business process problems to be addressed, the *Problem to Ideal Solution* phase consists in the following tasks:

- Class of problems 1 or 2: finding physical solutions for process reorganization and resources allocation.
- Class of problems 3: finding physical solutions for implementing the new product profile.

In the following subsections some references about appropriate methodological approaches are provided in order to guide the reader in the selection of the most suitable instruments to support the aforementioned activities.

2.2.4.1 Finding Physical Solutions for New Process Implementation

The value indexes, extracted as seen in Sect. 2.2.3.1, address the patterns for the overall reengineering of each phase of the business process. As already recalled, the directions to be followed can be classified in three main categories:

(1) Increasing the phase value through the improvement of its performance or in terms of efficiency, i.e. through the reduction of the involved resources, while preserving the same benefits. Such objective is classically pursued by technological enhancements, more efficient organization systems, broader employment of ICT to optimize the flow of resources within the process.
(2) Increasing the phase value by supplying new customer requirements. The scope can be attained by exploiting partially used resources or by-products in fruitful ways, capable to head towards the generation of additional features. With such aim, the business process model represents a suitable starting point for the individuation of not fully exploited resources.
(3) Suppressing low value phases, with the consequent modification of other process sections which are the candidates for the fulfillment of the consequently unsupplied customer requirements. In order to perform the task, it is useful to highlight further phases employing similar kinds of resources, technologies, know-how.

According to the above objectives, the authors put forward a set of acknowledged methodologies, aimed at addressing the task of identifying conceptual solutions.

Classical TRIZ tools, e.g. the *76 Standard Solutions* [11], represent suitable instruments to increase the performance or the efficiency of the phases (directions 1 and 2). More precisely, once the critical function of the phase to be enhanced has been identified, the Standard Solutions constitute general strategies to increase its effectiveness, through the introduction or modification of appropriate substances and/or fields (standards belonging to class 1.1) or through a more efficient use of the existing resources (standards belonging to class 2).

Many methodologies deal with policies within manufacturing environments and they are mostly tailored to reduce useless resources, so that they address the first direction for phases modifications. In this context, *Lean Manufacturing* [14] and *Quick Response Manufacturing* (QRM) [15] provide valuable suggestions for business improvements. Lean Manufacturing proposes a large set of tools that aim at reducing wastes, meant as those activities carried during the production stages that do not bring any added value. Lean Manufacturing introduces a pull-based supply chain, whereas procurement and production are demand driven and thus coordinated by actual customer orders. The supplying and the purchases are ruled by *Just in Time* (JIT) strategy that aims primarily at the reduction of in-process inventory. Besides, the reduction of the operational times can be obtained through the means of QRM, whose target is the minimization of lead-times. In order to provide further benefits, QRM methodology should be applied to the whole supply chain, strengthening the cooperation among the involved business units that participate in the generation of the value.

The assignation of new properties to a certain phase can be supported by the individuation of existing techniques in dedicated knowledge bases. Scientific documents and especially patents represent the widest available source of technical information close to the technological frontier. The individuation of proper ways to put in practice additional features of the phases can be done also with function retrieval tools. In the scope of TRIZ, *Function-Oriented Search* (FOS) [16] is especially suitable to find and apply existing functions, also from different technical fields. FOS is an evolution of the TRIZ concept assessing that the shortest path to an effective solution is to use an analogy. The tool leads the user in the identification of the key problem, the formulation of a generalized function to be achieved, the individuation of the most appropriate industrial area to be investigated, the selection of the technologies closest to required functional parameters.

2.2.4.2 Finding Physical Solutions for New Product Implementation

According to the results coming from the previous step of IPPR, a new set of product specifications in terms of value attributes is obtained. Thus, before performing any conceptual design activity, such attributes must be translated in candidate Engineering Requirements (ERs) of the new system. Among the others, a useful method used to support the preparation of the ERs list is the QFD, that helps

Table 2.3 The chart summarizes the flow of activities foreseen by IPPR for each class of BPR problems to be addressed

Phase	IPPR activity	Class of problems 1 and 2	Class of problems 3
Step 1			
Process to problem	Process modelling	•	
	Product information elicitation	•	•
	Product/service modeling	•	•
Step 2			
Problem to ideal solution	Identification of what should be changed in the process	•	
	Identification of what should be changed in the product/service		•
Step 3			
Ideal solution to physical solution	Finding physical solutions for new process implementation	•	
	Finding physical solutions for new product/ service implementation		•

to translate customer wants into product requirements. Moreover, through the QFD, the designer can have a clear vision of the criticalities related to design problem, since these tools allow the identification of any positive or negative correlation among the product requirements. Along the translation of customer requirements into engineering specifications, an iterative process is common to refine both lists, e.g. by highlighting possible new advantages arising by the profile conceptualization or the emergence of (at least apparently) mutually not compatible demands.

According to the nature of the design problem and its complexity degree, it may happen that no inventive step is required to obtain the successful solution, but just the application of the knowledge already available within the design team. The recalled TRIZ *76 Standard Solutions* are an excellent structured checklist which allows to browse the team knowledge with a systematic approach. Alternative methods to support this kind of design task are presented in [17].

Besides, if the previous analysis points to the necessity to overcome the emergence of conflicting requirements, the design task requires the application of tools for the identification and solution of contradictions, such as the techniques suggested by the TRIZ [11]. As a result, a conceptual solution is generated in terms of physical properties of the system that allows to satisfy the conflicting requirements according to the available resources.

2.3 Summary of IPPR Flow of Activities

The flow of activities foreseen by IPPR to address the problems of classes 1, 2 and 3, is summarized in Table 2.3, according to what is reported in the previous Section.

The reader can refer to this chart in order to easily identify the relevant tasks involved in each step of the method, that are required to address the faced reengineering problem.

References

1. Houben G, Lenie K, Vanhoof K (1999) A knowledge-based SWOT-analysis system as an instrument for strategic planning in small and medium sized enterprises. Decis Support Syst 26(2):125–135
2. Roth RM, Wood WC (1990) A Delphi approach to acquiring knowledge from single and multiple experts. In: ACM SIGBDP conference on trends and directions in expert systems, Orlando, 31 October–2 November 1990
3. Polanyi M (1966) The tacit dimension 1966. Anchor Day, New York
4. Nonaka I (1998) The knowledge-creating company. In: Neef D, Siesfeld GA, Cefola J (eds) The economic impact of knowledge 1998. Butterworth-Heinemann, New York
5. Mazur G (1997) Voice of customer analysis: a modern system of front-end QFD tools, with case studies. In: American society for quality conference, Orlando, 5–7 May 1997
6. Tontini G (2003) Deployment of customer needs in the QFD using a modified Kano model. J Acad Bus Econ 2(1):103–113
7. Curtis B, Kellner MI, Over J (1992) Process modelling. Commun ACM 35(9):75–90
8. Giaglis GM (2001) A taxonomy of business process modelling and information systems modelling techniques. Int J Flex Manuf Syst 13(2):209–228
9. Porter ME (1985) Competitive advantage. Free Press, New York
10. Kano KH, Hinterhuber HH, Bailon F, Sauerwein E (1984) How to delight your customers. J Prod Brand Manag 5(2):6–17
11. Altshuller GS (1984) Creavity as an exact science. In: Williams A (trans) The theory of solution of inventive problems. Gordon and Breach Science Publishers, New York
12. Miles LD (1949) How to cut costs with value analysis. McGraw Hill, New York
13. Mittal V, Ross WT, Baldasare PM (1998) The asymmetric impact of negative and positive attribute-level performance on overall satisfaction and repurchase intentions. J Mark 33(4):271–277
14. Womack J, Jones D (1996) Lean thinking. Simon and Schuster, New York
15. Suri R (1998) Quick response manufacturing: a companywide approach to reducing lead times. Productivity Press, Portland
16. Litvin SS (2004) New TRIZ-based tool—function-oriented search. In: Proceedings of the 4th TRIZ future conference, Florence, 2–5 November 2004
17. Pahl G, Beitz W (2007) Engineering design: a systematic approach, 3rd edn. Springer, London

Chapter 3
IPPR Implementation

3.1 Introduction

This Chapter illustrates the instruments for the implementation of the IPPR activities according to the methodology workflow described in Chap. 2. The instruments involved in each step of IPPR are hereby presented with the aim of clarifying their tailored application according to the three classes of reengineering problems.

The content is structured in three main Sections which comply with the steps of IPPR. Section 3.2 describes the tools that are suggested to carry out the activities foreseen by the *Process to Problem* phase. They deal with the process modeling, the information elicitation about the output of the business which is valued by the customer and the product representation. Then, with reference to the tasks to be performed in the *Problem to Ideal solution* step, Sect. 3.3 shows the instruments that are recommended to analyze the process and the product attributes, as well as to identify the appropriate directions to advantageously reengineer the system in the customer value perspective. Eventually, Sect. 3.4 provides the metrics to select the specific tools which are viable to support the redesign tasks, as foreseen by the *Ideal Solution to Physical solution phase*.

Table 3.1 provides a summary of the techniques proposed for each activity of IPPR, as well as the matching Sections. Thanks to the Table, the user can easily individuate the content which is individually deemed to be the most interesting and useful for the purpose of the reengineering problem to be faced.

3.2 Implementation of the "Process to Problem" Phase

The *Process to Problem* phase includes a set of activities whose outcomes are crucial for the subsequent analysis tasks. As recalled in Table 3.1, they consist in:

F. Rotini et al., *Re-engineering of Products and Processes*,
Springer Series in Advanced Manufacturing, DOI: 10.1007/978-1-4471-4017-7_3,
© Springer-Verlag London 2012

Table 3.1 The chart summarizes the tools required to perform the activities involved in each step of the IPPR methodology, according to the class of reengineering problems to be addressed

Phase	Activity	Class 1	Class 2	Class 3	Section
Step 1					
Process to problem	Process modelling	Multi-domain modeling technique			3.2
					3.2.1
	Product information elicitation	CRs checklist, correlation coefficients			3.2.2.1
				Lifecycle system operator	3.2.2.2
	Product modeling	Relevance scale			3.2.3.1
			Kano model		3.2.3.2
				Functional features	3.2.3.3
Step 2					
Problem to ideal solution	Identification of what should be changed in the process	Phase overall satisfaction			3.3
					3.3.1.1
			Customer satisfaction/dissatisfaction, phase overall satisfaction		3.3.1.2
		Resources consumption			3.3.1.3
		Overall value, POS–RES chart			3.3.1.4
			Phase values, value assessment chart		3.3.1.5
	Identification of what should be changed in the product			New value proposition guidelines	3.3.2
Step 3					
Ideal solution to physical solution	Finding physical solutions for new process implementation	Guidelines to select process redesign tools			3.4
					3.4.1
	Finding physical solutions for new product implementation			Guidelines to select product redesign tools	3.4.2

The last column shows the sections of the chapter where the user can find the detailed description of the instruments

- *process modeling* is required only for the problems related to classes 1 and 2 and it is aimed at obtaining an overall representation of the business process, capable to summarize the performed functions and the involved resources, the control parameters governing the phases, the employed technologies;
- *product information elicitation* is an activity aimed at collecting the relevant performances concerning the outputs of the business process for the classes of reengineering problems 1 and 2. In order to address the reengineering problems belonging to class 3, the information to be preliminarily organized concerns the value supposed to be delivered by the product within the whole lifecycle;
- *product modeling*: in the case of problems belonging to class 1 it is employed to assess the relevance of each customer requirement in generating the customer perceived value. With reference to the second class of reengineering problems, additional indications are required, concerning the role of the customer requirements in avoiding the buyer discontentment and/or providing an unexpected level of satisfaction. Conversely, when taking into consideration the problems belonging to class 3, the model to be used highlights the features offered to the customer, the ways such attributes deliver value and their performance levels.

Section 3.2.1 provides the description of the tools to support the information gathering, Sect. 3.2.2 illustrates the instruments employed to model the industrial process and eventually Sect. 3.2.3 describes the tools dedicated to the product modeling.

3.2.1 Multi-Domain Process Modeling Technique for Classes of Problems #1 and #2

The process modeling is a fundamental activity in order to address the reengineering problems belonging to the class 1 and 2, since its outcomes constitute the pillar on which the subsequent analyses are grounded.

As widely discussed in the previous Chapter, IPPR requires an exhaustive representation of the industrial process from both the technical and economic perspectives. To this end, the adopted technique merges together the flows of inputs/outputs relevant for the process analysis, classically represented through *IDEF0, Energy Material Signal* (EMS) and *Theory Of Constraints* (TOC) models. Such a schema allows the organization of a comprehensive set of information related to technical and economical domains in an overall frame capable to provide a detailed representation of the resources and monetary flows, expected outcomes of the phases, control parameters, involved technologies, etc.

The multi-domain technique is here presented focusing the attention on the features which are interesting for the process modeling within the context of IPPR, however the user can refer to the description of the IDEF0, EMS and TOC models reported respectively in the Appendixes A, B and C, to deepen his/her knowledge about the original scopes of these techniques.

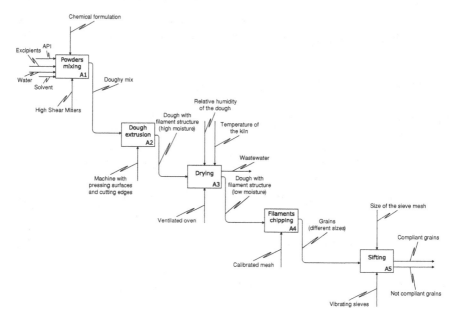

Fig. 3.1 Exemplary application of IDEF0 to model a pharmaceutical manufacturing process

IDEF0 is used to represent the constituent activities of the process and the conversion of inputs into outputs, in addition to the controls governing the transformation and the technologies required for performing the process. By employing the IDEF0 model, a preliminary schematization of the industrial activities and operations leading to the delivery of the product is viable to segment the business process into the phases to be subsequently analyzed. Indeed, coherently with the IDEF0, the business process is favorably represented within IPPR as a technical system constituted by chains of operations, whereas each box represents a phase. An exemplary application of the IDEF0 framework, with reference to an established manufacturing practice adopted in the pharmaceutical industry, is illustrated in Fig. 3.1, with the aim of showing the capabilities of the model to segment the process into the constituent phases.

The logic of the EMS model is exploited in order to consider also the flows of energy and signal/information among the involved resources. Suitable conventions can be adopted to remark such kinds of flows, as showed in Fig. 3.2, whereas the basic IDEF0 scheme for each single phase is enriched by the pertinent mapped features.

Eventually, the customized IPPR model is integrated with the highlighting of the monetary flows involved in the business process. The task requires therefore the indication of the expenditures involved within the system, to be favorably monitored throughout the terms that contribute to determine the Inventory and the Operating Expense, according to TOC model. In such a way, each activity can be characterized not only in terms of the parameters related to the technical domain, but also with different factors that participate to the generation of the costs to be

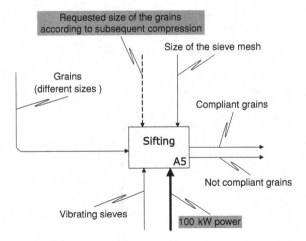

Fig. 3.2 The EMS model within the multi-domain model is adopted to represent the flows of energy, material and signal/information

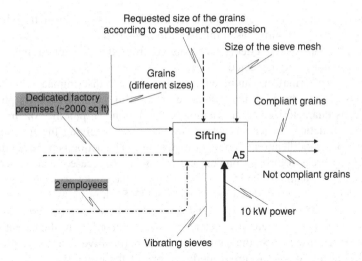

Fig. 3.3 Issues of the TOC model are employed to represent the flows of expenditures involved in the process

quantified for each phase: consumption and maintenance of the tools, labor, investments, expenditures for the plant or the offices, etc. The channeled resources, not previously mapped, that are due to Inventory and Operating Expenses, can be represented in the model through suitable conventions, as illustrated in Fig. 3.3, optionally indicating the amount of generated expenses.

As showed in Fig. 3.4, the model adopted for the purpose of IPPR requires to introduce the time needed to perform each phase, which is disregarded by the

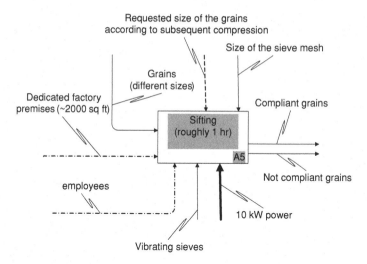

Fig. 3.4 Also the elapsed time of each activity is taken into consideration and represented in the model

previous techniques. The latter represents a relevant issue, especially when the competition is significantly based on the time to market.

Finally, the undesired effects that emerge from the display of the process activities are not to be neglected.

An enhanced formalism can be employed in order to discriminate when a flow or an action should be considered positive or harmful. Each elementary phase of the model is characterized by a function (or commonly a plurality of functions) involving a function carrier, an action and an object receiving the function and undergoing the modification of some of its features. The action is properly defined if it can be expressed through one among four verbs (increase, decrease, change, stabilize) and the name of the impacted property of the object, as suggested by the Element-Name-Value (ENV) model detailed in [1]. Such property (e.g. the length, the color, the electrical conductivity, the shape), is thus set to a certain value (e.g. one meter, red, five Siemens per meter, spherical), according to the extent of the function. As a consequence, the nature of the action (positive or negative) depends on the desirability of the occurred modifications of the property.

Diffusedly, unwanted phenomena associated with the phases result in additional expenditures for the firm in the form of introduced auxiliary operations (e.g. activation of security devices, employment of noise or emissions abatement equipment, machinery cleaning).

Thus, the overall consideration of involved resources has to include the undesired effects, currently beyond all remedy, and the auxiliary functions aimed at removing or mitigating the bad consequences of unwanted phenomena. Appropriate formalisms, as those proposed in Fig. 3.5, can be employed to describe such phenomena.

According to the description performed so far, the conventions adopted to build the Multi-domain process model are summarized in Fig. 3.6. Eventually, Fig. 3.7 depicts

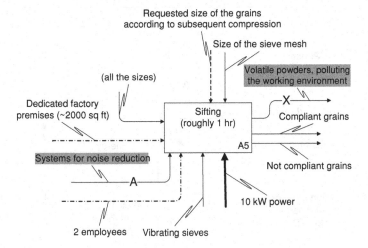

Fig. 3.5 The multi-domain model adopted by IPPR considers also the harmful or undesired flows involved in the phases, as well as the auxiliary functions needed to eliminate or mitigate unwanted phenomena

an example of graphical representation of a process phase according to the Multi-domain model.

The user is however free to adopt other graphical representations, techniques or models. Nevertheless, in order to perform the subsequent tasks, the process has to be structured by a list of industrial phases. For each of them the following information has to be gathered:

- inputs and outputs of materials;
- the energy and information channeled;
- the technologies to be adopted, the space occupied;
- the labor and personnel employed to perform, control and design the foreseen functions;
- the costs accounted in order to acquire or purchase the previously listed resources, as well as the financial investments necessary for the space occupied within the plant or the building;
- the elapsed time;
- drawbacks, undesired effects, auxiliary functions to face unwanted phenomena.

3.2.2 Tools for Product Information Elicitation

The application of IPPR to the reengineering problems of the classes 1 and 2 requires the identification of the value elements characterizing the product, expressed in the form of customer requirements. On the other hand, the fulfillment of a NVP task through IPPR, which is the activity performed to solve the

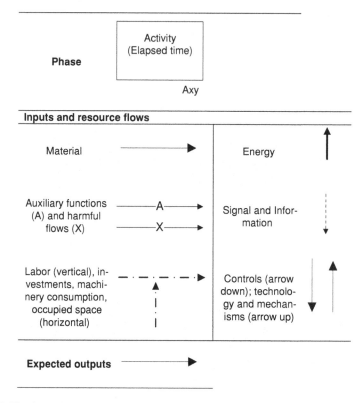

Fig. 3.6 The formalisms adopted to build the multi-domain model of the industrial process

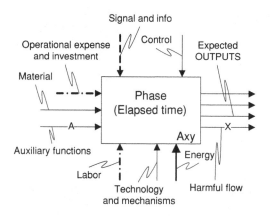

Fig. 3.7 Example of phase modeling according to the multi-domain technique

reengineering problems falling into class 3, needs the preliminary determination of the value elements offered by the as-is product, expressed in terms of value attributes. Thus, the investigation of the value elements held by the product/service constitutes a common task whatever the reengineering problem is. However, we have to highlight the following distinction:

- within the class 1, the customer requirements to be appointed are primarily (but not uniquely) those to be fulfilled within the process to be reengineered, as a result of underperformances or turbulences in the marketplace or in the boundary conditions;
- within the class 2, the customer requirements to be monitored are those delivered by the business process, as a result of the display of its phases;
- within the class 3, it is useful to map all the situations and circumstances influencing the employment of the product, its behavior, the drawbacks emerging as a result of its use or ownership. Thus the information to be collected and structured includes not only the explicit attributes, but also the complete experience generated along the product lifecycle, highlighting considered and unexploited sources of value and thus further business opportunities.

Then, whereas in the process reengineering tasks the customer requirements to be analyzed emerge as a result of the fulfillment of the phases, the mapping of the product attributes within the third class of business problems is advantageously enriched by the representation of the user experience, highlighting pros and cons.

This part of the Chapter offers the description of the tools suggested by IPPR to support the collection of data related to the industrial process and its outputs, i.e. *CRs checklist* (Sect. 3.2.2.1). Additionally, *Lifecycle System Operator* provides an useful aid to individuate the attributes of the product and the sources of value for the customer (Sect. 3.2.2.2).

3.2.2.1 Product Information Elicitation for Classes of Problems #1 and #2: CRs Checklist and Correlation Coefficients

The purpose of this activity is to determine a comprehensive list of customer requirements that participate to the generation of value and to measure the extent at which the process phases combine to bring about the fulfillment of the identified product attributes.

In order to consider the value aspects of the process outputs, a distinction can be made between processes operating in Business to Customer (B2C) or Business to Business (B2B) industries. In the first case the requirements to be satisfied concern the elements of value that can be appreciated by the mass of customers or a particular segment of end users. In the second case, the attributes to be taken into account regard both:

- the dimension of the industrial level of the direct customer, thus features such as the volume and the assortment of manufactured batches or additional services (e.g. transportation, certification, accomplishment of bureaucracies);
- the dimension of the quality aspects that result relevant downstream the supply chain, including the end user, which are enabled by a correct display of the analyzed business process; for example, a firm that processes raw copper and

manufactures wires has to fulfill certain requirements of the semi-finished products (size, flexibility, strength, etc.), which will allow an assembly industry to produce compliant electrical cables, further on appreciated by the end users.

It is worth to highlight that the elicitation of customer requirements can be carried out by considering different detail levels, leading to very diverging records of product attributes characterizing the outputs of the process. The same phenomenon regards the determination of the process phases. According to the authors' experience, in order to carry out equivalent analyses of the process and the product, it is recommended to use quantities of customer requirements and phases, whose ratio ranges from one half to two. Whereas such condition is not met, it is suggested to group the more numerous items within more general categories or to further segment the less abundant items (CRs or process phases).

In order to strengthen the elicitation of the attributes, IPPR proposes a tool, aiming at creating an exhaustive list of customer requirements. The proposed technique, namely *CRs checklist*, is a record of hints tailored to elicit the widest diffused kinds of product attributes. The newly individuated requirements are therefore to be grouped together with those extracted through the process analysis. CRs checklist recommends to consider a wider amount of issues, in order to identify further performances currently delivered to customers and stakeholders concerning the product and/or the offered batch of products, as suggested in [2]. The aspects to be considered are listed according to the *functional features* classification criteria, as follows:

- the useful functions (UF attributes), meant as the direct benefits perceived by the end user as a result of the product employment and more specifically:
  - the advantages arising from the exploitation of the product, which can be referred to the quality and the quantity of the desired output;
  - the amount of users for whom such benefits are met, thus the flexibility of the product according to different customer demands;
  - the capability of the product to meet the customer needs within the requested time;
  - the adaptability of the product when working in diverging conditions with respect to the designed preferred ones;
  - the stability of the product performances when subjected to external perturbations;
  - the chance to effectively control the system in order to obtain the expected outcomes;
  - the possibility to expand or upgrade the range of product functioning;
  - the opportunity provided to advantageously employ the product for not standard users or disabled people;
  - the possibility to customize the product or certain properties according to the user tastes and tendencies;
  - the possibility to use the system for different employments after the termination of main product functioning, the collection of matching items;

- the aesthetical requirements and the emotional dimension of the product, the style, the fashion content, what it evokes in the user, the lifestyle that the object implies, the prestige it generates for the owner as a feeling of distinction and recognition;
- the fun and adventure resulting from the use of the system;

• the strategies aiming at eliminating or attenuating the undesired effects (HF attributes), commonly associated with the product working:

- the integrity of the product itself, its resistance to planned or accidental stress or collisions, the strength against wear or corrosion;
- the limitation of damages towards treated objects or neighboring systems;
- the environmental sustainability, the recyclability, the possibility to reuse the system or its parts reducing the amount of waste;
- the ethics of the product as a distinguishing factor;
- the safety and innocuousness for human health and people's psychological and social conditions;
- the absence of bother for the user employing the product or for surrounding people, the comfort of use, the ergonomics, the manageability;
- the reliability, the limited frequency of system failures;
- the duration, the expected life of the product;

• the properties leading to the attenuation of the resources to be channeled by the buyer or the end-user of the system (RES attributes) and more specifically:

- the limitation of occupied space, the lessening of the encumbrance, the accessibility, meant as a shrunk quantity of space required to allow the users to employ, store, transport, maintain and dismantle the product;
- the working speed, the reduction of time to be waited before the functioning of the product delivers the expected outcomes, including the duration of the period to be waited before physically benefiting of the bought item or service after the purchase;
- the limitation of the time required to maintain or fix the product, to change accessories, to dismantle the system, to learn how to use it, to administer or to accomplish the involved bureaucracies;
- the reduction of the information and skills to be gathered in order to correctly use and control the product, the ease of employment, the user friendliness, the limitation of required training;
- the ease of acquiring the product, due to market penetration and distribution policies;
- the ease of managing, maintaining, assembling, disassembling, upgrading, substituting components or accessories;
- the ease of choosing and individuating the product in the marketplace, according to recognizable features, due to technical, aesthetical or communication issues;
- the lightness and the portability;

- the independence from the use of different materials, instruments, technical systems;
- the absence or limitation of the consumption of consumable items or materials;
- the reduction of auxiliary functions to be delivered in order to use, install, dismount or dispose the system;
- the limitation of the required energy needed for the product working, maintaining, installing, disposing, recycling; its efficiency;
- the decrease of the human power needed to use or transport the product;
- the additional services provided in order to attenuate the consumption of individual resources, as those listed in the previous bullets, the customer care.

The fact that the cost of the product for the user is not considered as a customer requirement, should not amaze, since, for the first two IPPR classes, it represents a direct consequence of the business process, the employment of the resources along its phases, as well as the pricing policies of the company.

As anticipated in Chap. 2, the process model eases the individuation of the main product performances that are intended to be fulfilled, by examining how the phases transform the inputs into the outputs. The strategy of exploring the performed process schematization can be used complementarily or in alternative to the employment of the CRs checklist. The model of the business process aids the elicitation of the customer requirements, by addressing for each process segment the question: "within the perspective of value delivery, which reason or scope motivates the transformation of the inputs into the outputs along the analyzed phase?". For example, if a process segment operates in order to modify the color of a certain object or material, IPPR users have to individuate the ultimate goal or the plurality of objectives to be achieved through the transformation. This leads therefore to determine the requirements to be fulfilled which are influenced by the color alteration, addressing to pertinent attributes, such as (according to the specific case) aesthetics, respect of norms, benefits concerning the heat transfer rate, intuitiveness of use, etc.

At the same time, the accomplishment of the customer requirements can be achieved by dealing with a set of parameters which are modified along the process display. Still by way of example, the designed style of a product, complying with the aesthetical requirements, can be attained by modifying both the color and the shape of an object. In simpler situations the fulfillment of customer requirements arises as a consequence of the modification or stabilization of a single parameter. In these cases there is a complete equivalence between the technical feature to be set and the customer requirement. Therefore, the distinction between an engineering characteristic and the met product attribute loses its meaning within the scope of IPPR.

Further on, the transformation of each parameter can be governed by one or more industrial activities, e.g. the correct dimension of the grains constituted by a pharmaceutical mixture is achieved by extruding and chipping the material, as suggested in Fig. 3.1.

In the context of product development, QFD [3] entails the identification of customer expectations and the rate at which engineering characteristics contribute to meet these needs. With a similar approach, in the scope of IPPR, the proposed procedure requires mapping the phases underlying the accomplishment of each CR and eventually evaluating the extent at which they participate to its fulfillment. Such contributions are designated with the name *correlation coefficients*. The book indicates each of these indexes with k_{ij}, meaning the relative contributions addressed to the j-th phase (within the record of process segments emerging from the process modeling) in ensuring the achievement of the i-th CR (according to the list of attributes resulting from the product investigation).

As a result of the concurrence of the (hypothetical) plurality of both modified parameters and industrial operations involved in the requirements fulfillment, the transformation chain implies that the attainment of each product attribute can result as the combined effect of more phases. In other terms, each process phase can contribute totally, partially or in no way to the delivery of satisfaction according to a specific customer requirement. The value assumed by the correlation coefficients, ranges therefore from 0 to 1; in each case the summation of the k_{ij} terms with respect to the list of phases has to be 1, as follows:

$$\sum_j k_{ij} = 1$$

It is worth to notice that each phase of the process model, if schematized according to the proposed framework, includes the parameters that govern and control the progress of the phases. The indicated issues favor the individuation of the mechanisms that lead to the accomplishment of the CRs, and thus the estimation of the correlation coefficients.

Further on, in order to correctly determine the k_{ij} indexes the analyzer has to take into account the extent of the transformations performed by the phases that lead towards the fulfillment of the CRs, as an immediate result (Fig. 3.8a) or by means of the accomplishment of engineering characteristics (Fig. 3.8b).

The recalled modifications that occur along the display of the process can regard both quantitative and qualitative parameters, to be expressed in terms of customer requirements or engineering features. In the first case the ratios measuring the influence on a parameter are fixed through a mathematical calculation, by considering how much the phase modifies that property with respect to the overall transformation that happens in the whole process. Otherwise, when dealing with qualitative aspects, the determination of the rates has to be carried out through estimations provided by sector experts. The same mechanism involves the measure of the impact of the engineering characteristics on the product attributes.

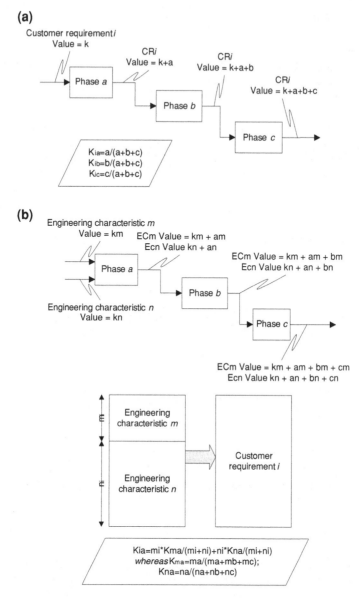

Fig. 3.8 Rules for the determination of the coefficients correlating the phases to the customer requirements. In **a** is shown the case in which the phases lead towards the fulfillment of the CRs, as an immediate result. While in **b** the case is presented where the CRs are obtained by the means of the accomplishment of engineering characteristics

3.2.2.2 Product Information Elicitation for Class of Problems #3: Lifecycle System Operator

With the aim of taking into account all the aspects that participate to the delivery of the value for the customer, the analysis has to be carried out by considering the following elements:

- *the life cycle*: by remarking any circumstance that may occur along the different stages of the existence of the product;
- *different levels of detail*: in order to pinpoint the potential benefits for the customer, that are likely to emerge by appropriately design the systems, at different hierarchical levels, that impact the product under investigation under various operating contexts.

Given its flexibility of use, the *System Operator*, as developed within TRIZ [4], can be employed as a powerful reasoning tool for mapping a wide range of situations, circumstances and working conditions otherwise neglected. In this context, a tailored version, namely *Lifecycle System Operator*, is used to elicit product attributes and individuate overlooked sources of value for the end-user. However, since the user can benefit of the original tool (beyond the customized one) in other contexts, he/she can find a comprehensive description of its objectives and characteristics in the Appendix D.

It is hereby proposed to adopt an appropriate subdivision of the temporal dimension of the product lifecycle, designed to pinpoint the most common situations in which the buyer can perceive value. Thus the time abscissa is considered starting from the moment the user begins to interact with the product. The lifecycle segmentation considers therefore the following phases:

- the purchasing, choice and access activities (e.g. ways of buying, determinants to select the product and its accessories, possibility to opt for a certain embodiment among different variants, negotiation for obtaining a service, fidelity bonds with the seller, trust in a trademark, awareness of the quality of the bought item);
- the operations and conditions preceding the employment of the system (e.g. mounting of an object, need of training, installation, suitable preparation of auxiliary systems, acquisition of documents, certificates or licenses);
- the utilization time, i.e. revealed performances of the system not previously evaluated, whatever impacts its employment, whatever allows it to function, the immediate impact of the working of the product on the surroundings and on the environment;
- the elapsed period before (and between) further exploitations (e.g. setup for a new employments, modification of the settings, keeping and maintenance, acquisition of novel functions, replacement of consumable items), as well as the impact of a single or a plurality of utilizations (e.g. health consequences, performance shrinking due to obsolescence);
- the phases related to the definitive termination of the functions, the dismantling (e.g. environmental issues related to the product disposal, recyclability,

reusability, alternative employments, collection of old items, negotiation to obtain new products or services).

Instead, in order to support the search of value elements according to different levels of detail at which the analyzed system can be considered during its life, it is suggested to organize the product dimensions in three main areas:

- the environment in which the product is situated;
- product or service itself, the operative zone;
- parts, components and accessories.

In accordance with these criteria, the customized version of the System Operator is presented in the Table 3.2, while an exemplary application is illustrated in Table 3.3, indicating the sources of value for a computer mouse.

The investigation schema suggested by the Lifecycle System Operator compels the user to ask himself the following question:

Are there any circumstances occurring during the <life cycle phase> and concerning the <product dimension>, to be observed and treated, resulting as inputs for a valuable design of the product?

Thus, the Lifecycle System Operator can be employed as a collection of fifteen questions, which support the scope of systematically browsing the possible sources of value offered by the product.

Subsequently, the individuated sources of value have to be appropriately elaborated and interpreted in order to elicit product attributes. With the objective of expressing the benefits ascribable to the product design in terms of value attributes, the user has to characterize the competing factors as parameters, whose increase in the offering level results in enhanced customer satisfaction.

The paths leading to the definition of the attributes differ according to the way the value sources have been depicted within the Lifecycle System Operator. In some cases, the value sources already express the feature that should be mapped as a product attribute (e.g. *lightness* or *shock resistance* in the Table 3.3). In other circumstances, undesired conditions are mapped and they have to be uttered in such a fashion as to individuate the capability to limit the related inconveniences, drawbacks or resources channeling. For instance, with reference to Table 3.3, the *cleaning issues* can be appropriately expressed in terms of *ease of cleaning*, *quickness of cleaning*, *limited frequency of required cleanings*, *absence of the need to employ particular products to perform the cleaning*, *absence of the need to clean the mouse*, etc. Further on, the sources of value can be expressed in terms of main or additional useful functions: in these conditions, the analyzer has to point out the benefits that arise as a result of the display of the recalled functions (e.g. from *scroll-up* in Table 3.3 up to define the *capability to support the end-user in browsing the file*).

In any situation, the translation of the monitored value sources into customer requirements can be carried out through the following questions:

Table 3.2 Lifecycle system operator: a tailored tool for the investigation of the value elements

Product dimension	Lifecycle dimension				
	Purchasing, choice and access activities	Before use operations	Utilization time	Elapsed time before further exploitations	End of the functioning
Environment in which the product is situated	•	•	•	•	•
Product or service level	•	•	•	•	•
Parts, components and accessories	•	•	•	•	•

Table 3.3 Lifecycle system operator adopted to overview the sources of value and competitive advantage for a computer mouse

Product dimension		Lifecycle dimension				
		Purchasing, choice and access activities	Before use operations	Utilization time	Elapsed time before further exploitations	End of the functioning
Product dimension	Environment in which the product is situated	Availability of sellers (shops, stores, E-commerce, etc.)	Compatibility with different computer ports; Ease of installation; Configuration speed	Compatibility with operative systems; Energy consumption; Surface adaptability; Possibility to regulate the wire length; Absence of wire	Maintenance and reparability issues; Cleaning issues; Limitation of human damages (i.e. carpal tunnel syndrome)	Alternative employments in event of failure (i.e. collecting)
	Product or service level	Availability of shapes and colours; Cheapness; Design; Pointing functions (buttons, scroll-up, trackball, etc.); Other functions (i.e. light, calculator); Lightness; Overall environmental sustainability; Customization, originality; Mouse pad matching; Accessories and matching stuff	Supplied case for keeping the object	Manoeuvrability, ease of flowing; Grip, stability (i.e. against turning upside-down); Ergonomics, comfort of use; Pointing precision; Availability of touch surface for clicking; Shock resistance; Flexibility of use (i.e. maximum allowed distance from computer); Indications about expected duration margin; 3D usability	Possibility of rewinding wire; Ease of rewinding wire; Item durability; Battery duration; Reduced encumbrance; Transportability	
	Parts, components and accessories	Technology preference (mechanical, optical, laser)	Possibility to nest components into each other or to conveniently arrange for storage	LED guidance	Ease of batteries replacement; Cheapness of replaced batteries	Ease of disassembling; Components recyclability

- which objective(s) can be achieved when the product or the surrounding settings are designed to exploit the conditions emerging by considering the <value source> in order to positively impact on customer satisfaction?
- which property(ies), parameter(s) or performance(s), regardless it(they) is(are) currently fulfilled or not, is(are) meant to be introduced, incremented or stabilized in order to attain the <previously individuated objective(s)> or, however, to enhance the present situation?

An example is provided in the followings, by treating the value source *mouse pad matching*, still picked up from Table 3.3, and providing likely answers:

- Q: which objective(s) can be achieved when the product or the surrounding settings are designed to exploit the conditions emerging by considering the *mouse pad matching* in order to positively impact on customer satisfaction?
- A: *combination of the shape and the design of the mouse and the pad*
- Q: which property(ies), parameter(s) or performance(s), regardless it(they) is(are) currently fulfilled or not, is(are) meant to be introduced, incremented or stabilized in order to attain the *combination of the shape and the design of the mouse and the pad* or, however, to enhance the present situation?
- A: *overall aesthetics, fun, ease of making the mouse flow, ease of choosing an appropriate combined mouse pad.*

The so gathered sample of product attributes can be favorably integrated with any further neglected customer requirement. Additional features can be individuated throughout the *CRs checklist* and by considering not yet monitored costs, to be expressed in terms of the cheapness in performing the purchasing, the assistance, the maintenance or other services. For instance, by considering UF categories referable to the versatility of the product, an additional value attribute could be represented by the *possibility to be used by both right- and left-handed people.*

3.2.3 Product Modeling

The complete representation of the product as a collection of customer requirements does not include just the record of relevant attributes, but a further characterization, which depends on the kind of reengineering problem to be solved.

Within the class 1, any IPPR user has to indicate the importance of each single customer requirement within the value delivery. Such pointer is needed also for the second class, which however requires a twofold way of representation. For this kind of reengineering problems the distinction is performed between attributes capable to delight customers and basic requirements to avoid dissatisfaction. The employment of Kano model, reported in Appendix E, is recommended within the scope of IPPR. In order to carry out the above characterization of the attributes, the literature offers a wide coverage of applications that employ customer surveys for the determination of attributes relevance and role in the perspective of value

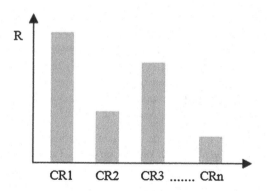

Fig. 3.9 The representation of the relevance indexes R for the set of customer requirements

delivery. Kano model itself has been developed as a customized strategy to extract and represent the Voice of the Customer. Along the time many alternatives have been proposed to indicate the most suitable CR accounted importance and Kano category, according to the outcomes of the surveys. IPPR users can employ any of the preferred models, by individuating the relevance indexes and distinguishing among Must-Be, One-Dimensional and Attractive features. However, such issues to be represented can be directly stated by business experts, whereas opinions of the clientele are unavailable or considered untrustworthy in the perspective of reengineering tasks, or the accomplishment of customer interviews is considered a too time-consuming task. The Sects. 3.2.3.1 and 3.2.3.2 provide a practical procedure to carry out the product modeling task, which is particularly suitable in event of the lack of customer surveyed opinions.

The classification of the product attributes for the scope of the third class is relevant in the perspective of the generation of a new product profile by the means of IPPR tools. The required characterization of the customer requirements in terms of functional features is described in the Sect. 3.2.3.3, which includes additional models to be further employed in order to ease the task of creating valuable profiles and choosing among alternatives.

3.2.3.1 Relevance Scale for the Classes 1 and 2

As previously recalled, the present activity deals with the determination of the extent (relevance index R) at which each customer requirement impacts the perceived satisfaction. Thus, the task is addressed at individuating, among the listed attributes, which features mostly impact the customer appreciation, motivate the choices among the products in the marketplace, result as a driver for promoting the buyer's loyalty.

Within IPPR it is suggested to express the relevance indexes with natural numbers through a Likert-type scale; in the performed IPPR implementations, including those presented in the Chaps. 4 and 5, the interval of scores for the R coefficients used to range from 1 to 5. By adopting this criterion the CRs char-acterized by a high relevance index relate to those competing factors playing a major influence within the customer experience when using the product. Conversely, the attributes characterized by low values of the R coefficient are assumed to be marginally relevant for the customer satisfaction.

Figure 3.9 reports a histogram which summarizes the extent of the importance indexes with reference to each customer requirement individuated in Sect. 3.2.2.1.

The notation R_i will be further on indicated to express the relevance of the generic i-th customer requirement.

3.2.3.2 Role of the Attributes Within Value Delivery for the Class 2

The hereby described task refers to the categorization of the customer requirements by considering their likely capability in providing unexpected value and/or guarding against strong discontentment. The logic of the classification to be per-formed follows the general idea of the Kano model, described with greater detail in the Appendix E. The introduction of different clusters for the product attributes is motivated by the need to individuate diversified directions of process reengineering with regards to the "kind of value" that is predominantly attained by the phases.

In order to establish the most suitable CR category, which describes the role in the determination of the perceived satisfaction, IPPR users has to answer the following questions for each listed product attribute:

- can the improper design and fulfillment of the <i-th CR> provoke customer dissatisfaction and rejection, since expected and demanded features are not met?
- can a correct accomplishment of the <i-th CR> combine to bring about cus-tomer appreciation, due to the generation of an unforeseen level of satisfaction from the buyer's viewpoint?

If just the first answer is affirmative, the investigated CR pertains the achievement of basic product characteristics and it can be classified within Must-Be (MB) attributes. If just the second reply is 'Yes' the related CR acts as a delighter for the consumer and has to be referred to Attractive (AT) competing factors. If both the answers are positive, the attainment of the analyzed CR results in the proportional delivery of perceived satisfaction according to the performance at which the attribute is provided, whereas low offering levels determine discontentment. Such require-ments deal with the category of One-Dimensional (OD) competing factors. Even-tually, if both the answers are 'No', the involved attribute does not provide any contribution in the product appreciation (Indifferent CR in the jargon of the Kano model) and it has to be crossed off from the list of relevant features. The logic of the attribution of the categories for the CRs is clarified through the Fig. 3.10.

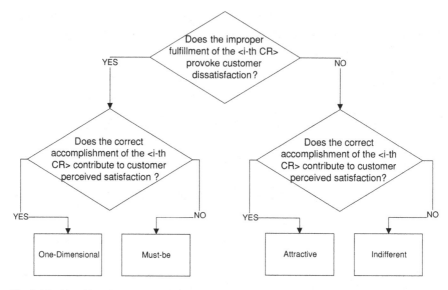

Fig. 3.10   Algorithm for the classification of the product attributes according to the Kano model

3.2.3.3 Categorization of the Attributes Through the Functional Features and Performance Evaluation for the Class 3

A step forward in the representation of the relevant product information for the third class of reengineering problems involves a twofold characterization of the collected attributes.

At first, a categorization has to be performed according to the clusters of *functional features* already illustrated in Sect. 3.2.2.1, by including the attributes pertaining the customer expenditures in the RES class. Each mapped customer requirement has to be therefore associated with the sound functional group, by considering from the buyer's viewpoint the benefits attained by the fulfillment of the attribute. If the task results complicated for the IPPR user, the following framework can result useful:

- if we consider the <product attribute>, are we dealing with the endeavor to request the customer less money, time, energy, space, tools, materials, information, experience or know-how? If the answer is YES, then the requirement can be addressed as a RES attribute. If the reply is NO, go to the following;
- if we consider the <product attribute>, are we dealing with the objective of reducing the impact of an undesired event, generally associated with the product functioning or decrementing the probability of such unwanted situation? If the answer is YES, then the requirement can be addressed as a HF attribute. If the reply is NO, go to the following;
- if we consider the <product attribute>, are we dealing with the effort of increasing the benefits for the customer or for a circumscribed group of users,

Table 3.4 Framework for summarizing the current product profile

Product attribute	Functional feature	Performance and customer demand
Lightness	RES	Good, outstripping the demand
Ease of cleaning	RES	Low, barely sufficient
Absence of the need to clean the mouse	HF	Absent, unsatisfying
Shock resistance	HF	Good, adequate
Capability to support the end-user in browsing the file	UF	Good, adequate
Overall aesthetics	UF	Moderate, barely sufficient

the versatility of the product functioning, the stability of the outcomes, the delight or happiness generated by the treated system? If the answer is YES, then the requirement can be addressed as a UF attribute. If the reply is NO, the attribute should be best deleted from the list in order to avoid subsequent bias.

In such a way the *lightness* and the *ease of cleaning* are, for instance, referred to RES attributes, since they work against a major involvement of the user individual resources and capabilities. Conversely, the *absence of the need to clean the mouse* and the *shock resistance* belong to the HF class, due to the intended scope of limiting the impact of undesired effects, such as the presence of dirty agents or any physical damage of the device. Eventually, the *capability to support the end-user in browsing the file* and the *overall aesthetics* fall into the cluster of UF customer requirements, since they are designed to directly give rise to benefits, accomplish requested tasks, arouse positive or playful emotions.

Finally, IPPR methodology requires the indication of the offering level that characterizes each product attribute, with a particular reference to customer exigencies, as arising by surveys or experts evaluations. Some features exactly meet the buyer expectations and needs, being the provided performance displayed at a degree capable to satisfy the customer, while higher levels would not result in increased contentment. In other cases, the desired performance degree of a certain attribute cannot be reached due to technological limitations or because of trade-offs with different conflicting demands. In further circumstances, the features result oversupplied, being their offering level greater than the actual requirements or expectations of the clientele. Generally speaking, the characterization of the offering level of the product attributes has to include both a qualitative level of the performance (e.g. absent, low, moderate, good, very high) and an evaluation about its relationship with the customer expectations (e.g. unsatisfying, barely sufficient, adequate, outstripping the demand).

Table 3.4 shows an exemplary classification of the previously cited product attributes pertaining a computer mouse.

It is worth to notice that the performance levels can strongly depend on different product configurations and variants. Hence, in case of treating products with fairly standardized value propositions, the presented scheme is sufficient to highlight the peculiarities of the marketplace. Conversely, if the distinction between two or more profile options is relevant within the aim of the reengineering initiative, the representation of the delivered product performance can be consolidated by employing the Value Curve (recalled in Chap. 1.3.2.3), a tool introduced within the Blue Ocean Strategy.

3.3 Implementation of the "Problem to Ideal Solution" Phase

This Section presents the instruments suggested by IPPR to perform the activities included in the Step 2 of the whole methodology. Depending on the reengineering problems, the main activities are:

- *class of problem* 1: identifying what should be changed in the process in order to overcome market boundaries;
- *class of problem* 2: identifying what should be changed in the process in order to recover competitiveness;
- *class of problem* 3: identifying what should be changed in the product in order to create a novel attracting profile.

Section 3.3.1 reports all the instruments required for the implementation of IPPR to address reengineering problems belonging to the class 1 and 2. Besides, Sect. 3.3.2 illustrates the tools needed to perform the NVP task which is expected to solve the reengineering problems of the class 3.

3.3.1 Performing the Identification of What Should be Changed in the Process

The instruments hereby presented support the execution of the activities involved in the identification of the process criticalities and the individuation of the most appropriate reengineering actions. These tools altogether allow:

- the determination of the *Phase Overall Satisfaction* index for the problems belonging to class 1 (Sect. 3.3.1.1);
- the calculation of the *Phase Overall Satisfaction* index for the problems belonging to class 2 (Sect. 3.3.1.2);
- the evaluation of the *Resources consumption* index for both the process oriented classes (Sect. 3.3.1.3);
- the assessment of the *Phase Overall Value* and the creation of the *PRAC* diagram for the classes 1 and 2 (Sect. 3.3.1.4);
- the determination of the *Value indexes* related to each process phase and the building of the *VAC* graph, pertaining the second class (Sect. 3.3.1.5).

The Phase Overall Satisfaction (POS) index represents a measure of the contribution that each process phase provides in determining the benefits perceived by the customer. The coefficient that assesses the Resources consumption is an overall measure of the investments and drawbacks faced by the company to carry out the process phases. The Overall Value (OV) index is viable to compare the benefits and the needed resources of each phase, thus elucidating which process segments are the most capable to generate satisfaction, according to the price paid by the company, and which ones result as bottlenecks in the same perspective. Further insights about the value delivery can be extracted throughout the diagrams named *POS versus RES Assessment Chart* (PRAC) and *Value Assessment Chart* (VAC), which characterize the nature of the criticalities, thus orientating the user towards suitable directions for process reengineering.

3.3.1.1 Measure of the Overall Satisfaction for Class 1

The schema adopted to evaluate the benefits for the customer, as they arise by the process, takes into account the impact of the phases in fulfilling the CRs, with a particular emphasis on the attributes characterized by a greater relevance. In this sense, the extent of the phases in the determination of the customer satisfaction is measured with regards to the combined effect of the contribution in delivering each requirement and the importance of the corresponding product feature in the perspective of value building.

As a result, the POS index for problems related to class 1 is calculated through the following expression:

$$POS_j = \sum_i k_{ij} \times R_i.$$

Thus, the esteem of provided satisfaction comes out by summing the shares of relevance indexes ascribable to the considered phase j.

3.3.1.2 Measure of the Overall Satisfaction and Other Indexes for Class 2

As illustrated in Sect. 3.2.3.2, the classification schema adopted by IPPR based on the Kano Model clusters the relevant customer requirements in three main categories that play a different role in the product perception: Must-Be, One-Dimensional and Attractive. On these bases, the class of problem 2 computes, for each CR, the terms expressing the capability to deliver customer satisfaction and indexes that measure the extent in avoiding discontentment. More specifically, as anticipated in Chap. 2:

- *Customer Satisfaction (CS)* represents the contribution given by an attribute to provide satisfaction with respect to the product or the service when the related CR is fulfilled;

Table 3.5 o_i, a_i and m_i coefficients according to the Kano classification of the i-th CR		Must-Be	One-dimensional	Attractive
	m_i	R_i	0	0
	o_i	0	R_i	0
	a_i	0	0	R_i

- *Customer Dissatisfaction (CD)* indicates the extent of the risks occurring when a given attribute of the product or the service is not met.

The above terms are calculated through expressions borrowed by literature contributions aiming at developing Kano model and extending its scope [5, 6]. In such a way, the CS and CD coefficients are calculated through the following expressions:

$$CS_i = \frac{o_i + a_i}{A + O + M}; \ CD_i = -\frac{m_i + o_i}{A + O + M};$$

where:

- CS_i and CD_i are respectively the Customer Satisfaction and Dissatisfaction indexes for the i-th CR;
- A, O and M are the sums of the relevance indexes of Attractive, One-Dimensional and Must-Be CRs along the whole record of product attributes; thus the following formulas apply:

$$A = \sum a_i; \ O = \sum o_i; \ M = \sum m_i$$

- o_i, a_i and m_i are equal to 0 or correspond to the relevance degrees of i-th CR depending on whether it is classified as One-Dimensional, Attractive or Must-Be, as summarized in the Table 3.5.

The previously calculated indexes CS and CD, referring to each relevant product feature, allow to compute the *Phase Customer Satisfaction* (PCS) and the *Phase Customer Dissatisfaction* (PCD) coefficients, which represent the contributions of the phase in determining unexpected appreciation and avoiding discontentment.

PCS and PCD are determined through the following relationships, which take into consideration the correlation coefficients and the extents of the requirements in producing unspoken benefits and guarding from dissatisfaction:

$$PCS_j = \sum_i k_{ij} \times CS_i; \ PCD_j = \sum_i k_{ij} \times CD_i.$$

As discussed in Chap. 2, the POS is assessed, within reengineering problems of the second class, through an empirical function [7] which combines into a non-linear expression satisfaction and dissatisfaction coefficients. As a result, the POS is calculated through the formula:

$$POS_j = 0.29 \times PCS_j - 0.04 \times PCS_j^2 - 0.72 \times PCD_j + 0.07 \times PCD_j^2.$$

3.3.1.3 Resources Consumption for Classes 1 and 2

The business process has to be characterized also through coefficients that measure the degree of the impact played by undesired issues which, in the company perspective, are associated with the display of the phases. As represented in the process model, the whole range of disadvantages is constituted by elapsed times, harmful phenomena and costs, determined by a wide variety of resources to be channeled and auxiliary operations that allow the execution of the phases. The *Resources consumption* index is aimed at quantifying the extent of such inconveniences occurring during each process phase. The coefficient is evaluated through the following formula:

$$RES_j = c \times C_j + t \times T_j + h \times HF_j$$

where:

- C_j represents the share of the total costs incurred to carry out the j-th phase (materials, energy investments, auxiliary functions, labor, space, machinery, etc.);
- T_j indicates the share of time spent in completing the j-th phase, with reference to the whole business process;
- HF_j is the share of the estimated damage produced by harmful effects arising from the j-th phase;
- c, t and h stand for coefficients, determined by business process experts, expressing the relevance of expenditures, elapsed times and drawbacks in hindering the market access (1st class of problems) or the preservation of the competitiveness (2nd class of problems).

The employment of shares instead of real values is dictated by the need to sum parameters with different units of measurement. The introduction of the lastly mentioned coefficients is proposed to take into account different situations, e.g. c is predominant when the business process is associated with a very poor profit, t grows when the time to market is a relevant competing factor, h assumes considerable values when the undesired aspects produce great drawbacks for the life and the image of the company.

In order to better compare the Resources consumption extents of the various phases, it is suggested to determine the normalized values of these coefficients. In this way the phases can be characterized by their accounted relevance in generating demands and undesired effects for the business process.

Fig. 3.11 The POS–RES
assessment chart. The
positioning of the process
phases according to the POS
and RES indexes, facilitates
the identification of the
process criticalities and of the
subsequent reengineering
actions

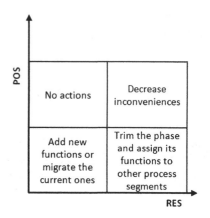

3.3.1.4 Overall Value and PRAC Diagram for Classes 1 and 2

The ratio between the general amount of benefits and the extent of disadvantages heads to the determination of the *Overall Value* (OV) coefficient. Such index, if referred to the j-th phase of the process, is calculated through the following relationship:

$$OV_j = \frac{POS_j}{RES_j}.$$

The extents of the OV parameters pertaining each phase can be advantageously normalized in order to express the global value of the process segments through percentage scores.

The OV indicator is suitable to identify, at a first evaluation level, strengths and weaknesses of the business process with reference to the set of phases. According to this metric, the phases showing a high OV rate can be considered to be tailored to the business process and their employed resources are well spent in generating customer satisfaction. Conversely, the process segments with low OV scores represent problematic issues and bottlenecks in the value creation process.

The separate consideration of POS and RES indexes is viable to highlight the nature of the bottleneck, by referring, more specifically, to a poor participation to value generation and/or to an excessive amount of expenditures and inconveniences. The above coefficients are employed to build the *POS versus RES Assessment Chart* (PRAC) (Fig. 3.11), which positions the phases in a diagram capable to illustrate the overall situation of the process with regards to the spent resources and the amount of benefits generated for the customer. Such diagram is thus capable to summarize relevant process aspects in terms of both the quality of the outputs and the internal demands through the introduction of ad-hoc metrics pertaining the performed business phases. In other words, the PRAC facilitates the analysis of the process criticalities by splitting the focus on the numerator of and on the denominator of the ratio expressing the OV.

According to the position assumed by each phase in the chart, it is possible to identify the most suitable directions for the business reengineering initiative, by aiming at removing the process shortcomings and bottlenecks. The redesign actions to be pursued depend on the order of magnitude of the POS and RES indexes and more in detail on the distinction between low and high values for the same coefficients. By using a practical criterion an index is conceived as low when its value is smaller than the mean assumed by the sample of phases (and high, obviously, in the opposite case). Alternatively, whereas the record of investigated phases is particularly rich (indicatively, more than 15 items), the low coefficients can be considered those whose values range in the first quadrant if the overall set of quantities is considered. The emerging decision support for BPR task is articulated as follows with reference to the values assumed by POS and RES:

- *Low POS and high RES*: the scarce performances of the phase are related to both a considerable consumption of resources and a low contribution in determining the customer satisfaction. The reengineering task should evaluate the opportunity to eliminate the phase by assigning its delivered benefits to another segment of the process.
- *Low POS and low RES*: the poor OV rate of the phase is due to a limited contribution to the customer satisfaction. The reengineering actions to be undertaken should be oriented towards assigning the treated phase new functions to be delivered without a meaningful increase of the needed resources. As an alternative, given the poor benefits, it could be evaluated whether it is possible to integrate the currently performed functions within the display of other phases.
- *High POS and High RES*: in case the phase assumes a low OV rate, this has to be ascribed to a high denominator, i.e. high expenses or drawbacks. The focus of the reengineering initiative should be addressed towards a reduction of the main cost factors. In this sense chance is constituted by the substitution of the technology adopted so far.
- *High POS and low RES*: in such a case the phase plays a relevant role in determining the customer contentment and it the employs a low amount of resources, thus it should not be subjected to any reengineering action.

3.3.1.5 Phases Value and VAC Diagram for Class 2

In addition to the OV index, other metrics can be likewise introduced to point out the relationship between provided benefits and spent resources. The task should be carried out by resorting to the coefficients PCS and PCD, which have been introduced in the Sect. 3.3.1.2, giving rise to the *Value for Exciting requirements* (VE_j) and the *Value for Needed requirements* (VN_j). The last parameters represent the suitability of the resources employed along the phases in achieving customer satisfaction through unexpected properties of the product (VE_j) and in fulfilling the basic requirements so to avoid strong consumer discontent (VN_j). They are calculated through the following formulas:

Fig. 3.12 The value
assessment chart. It allows
the clustering of the process
phases in four main areas
which are related to the
performances in determining
the customer satisfaction and
in avoiding the dissatisfaction

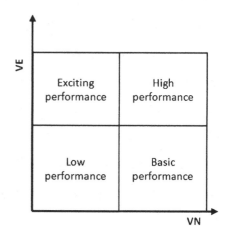

$$VE_j = \frac{PCS_j}{RES_j}; \; VN_j = \left|\frac{PCD_j}{RES_j}\right|.$$

In order to enhance the usability of the terms by employing more meaningful values, the VE and VN coefficients can be advantageously transformed into normalized extents or relative percentages.

A more intuitive analysis can be performed by considering the *Value Assessment Chart* (VAC) which is a useful representation of the phases according to the VE and VN indexes (Fig. 3.12). The values of VE and VN are considered low or high with the same criteria illustrated in Sect. 3.3.1.4.

As well as the PRAC, the VAC highlights the process criticalities and it leads towards the determination of the most appropriate reengineering directions for the removal of value bottlenecks. However the VAC adopts a different perspective than the first diagram, by focusing on whether the employed resources are well calibrated to guarantee the customer satisfaction and/or avoid dissatisfaction. The VAC graph can be used conjointly with the PRAC diagram or as an alternative for the decision support about the nature of the reengineering initiatives to be pursued.

According to VAC the process phases can be clustered in four main areas:

- *Low performance (low VN and low VE)*: the employed resources do not guarantee an adequate appreciation level of the product and they cannot avoid consumer dissatisfaction. The phases falling in this area thus need strong changes and also the opportunity of their elimination should be considered. It has to be investigated whether the low VE and VN rates depend on low benefits or high employed resources. A phase belonging to the former set is often worth to trim, by assigning the same minimal benefits to other existing phases. Besides, if the low value is due to high resources consumption, specific actions aimed at determining a leaner phase should be applied (indeed, this is the case when Lean Manufacturing provides maximum benefits). A further opportunity is to use the excess of resources for generating new attractive properties within the phase.

- *Basic performance (high VN and low VE)*: the employed resources do not provide unexpected benefits for the customer, but they are well spent to avoid consumer dissatisfaction. Typically, such phases are already optimized and oriented to fulfil the fundamental attributes; they do not need strong modifications and are not worth of consistent investments.
- *Exciting performance*: in this case, employed resources play an evident role to produce an adequate product appreciation level but they cannot avoid consumer dissatisfaction. Such phases are worth of investments in order to maximize their generated benefits; their success is a key to let the product to differ from the competitors.
- *High performance*: this quadrant is characterized by phases capable to provide well perceivable sometimes even unexpected benefits, still maintaining an extreme efficiency for fulfilling basic necessary needs. These phases are excellently tailored to the business process and they are worth to be safeguarded due do their high performances.

3.3.2 Performing the Identification of What Should be Changed in the Product

As inferable from Chap. 1.3.2.3, any reorganization of the product profile to be attained through a New Value Proposition (NVP) involves consistent modifications in terms of the value attributes offered to the customer and of their displayed performances. According to the scheme suggested by the Four Actions Framework (FAF, including Eliminate, Raise, Reduce, Create), the endeavor of a NVP task should be oriented towards introducing new competing factors and emphasizing those product attributes, whose offering level is still inadequate. Since such measures could go to the detriment of other valuable product aspects, it is recommended to miss out those customer requirements on which the market has long competed on or that result oversupplied.

By no way such ideal conditions can be encountered in any specific reengineering activity, technical field, industry or market. It can indeed happen that the customers of certain products, although in need of profound value redefinition, cannot give up (at least apparently) well established benefits. In different circumstances, showing a major need for the profile redesign, a big amount of alternatives can result viable on the basis of the indications provided throughout the FAF.

The tool that IPPR proposes for determination of an ideal product profile, still within the range of the solutions feasible at a first instance, fortifies the applicability of the Four Actions. The adopted framework, namely *New Value Proposition Guidelines* (NVPGs), supplements the mentioned reengineering actions with recommendations about which attributes to be subjected to the profile transformation process. The purpose of the suggested technique is to indicate what should be best

Table 3.6 Summary of the new value proposition guidelines	Most favorable actions	Actions to be maximally avoided
	Create RES	Reduce RES
	Create UF	Eliminate HF
	Raise HF	
	Raise RES	

done and should be avoided at a maximum extent, with regards to the functional features through which the product attributes have been classified. In other words, the guidelines remark which categories of competing factors are preferentially transformed within value transitions to be designed with respect to the Eliminate, Raise, Reduce, Create actions belonging to the FAF.

The NVPGs originate from an in-depth analysis of successful experiences, among which examples used as textbook cases for BOS, and stories of market flops, all concerning radical value transitions. The outcomes arise as a result of an initial part of the investigation, described in detail in [8, 9], and of further not yet published researches.

As a result of the way the guidelines have been extracted, they are structured as a collection of suggestions in terms of the functional typologies concerning the customer requirements to be involved in the transformation of the value profile. Hence, they individuate with major confidence the new valuable product attributes to be created, the existing properties to be enhanced, the current features whose performances are viable to be reduced and eventually the product characteristics to be eliminated without relevant drawbacks. The robustness of the arisen indications has been checked by the means of a χ^2 test, adopted to highlight whether the crossed distribution of actions and functional features could be due to chance.

The NVPGs, as resulting from the conducted survey, can be expressed as follows or, more schematically, through Table 3.6:

- *Create action*: considerable advantages arise by introducing neglected features, centered on the reduction of employed resources within the buyer perspective; a considerably positive role is played also by the emergence of novel function-alities or not previously considered characteristics impacting the user state of mind; the generation of new attributes aimed at facing unresolved troubles provides minor benefits;
- *Raise action*: it is observed that the meaningful mitigations of the inconve-niences due to Harmful Functions (HF) and to the consumption of Resources (RES) are the most recommendable; a leap concerning the performance of the functional requirements results in less evident advantages;
- *Reduce action*: while the drop of the performances of the attributes classified as UF and HF (hence the deterioration of the impact due to undesired phenomena) is tolerated, major drawbacks are caused by a considerable increase in the resources employment;

- *Eliminate action*: whereas the NVP task can bear the elimination from the bundle of product attributes of features clustered with UF or RES (thus the need to employ kinds of resources not previously engaged), the emergence of unprecedented undesired effects maximally contributes to market failure.

The so determined guidelines are viable to support the process of generating innovative product profiles or business models, by considerably delimitating the space for alternatives within new value proposition tasks. However, the indications dictated by the suggested tool have to be maximally harmonized with the mentioned general criteria involving the employment of the FAF. In other words, the designed actions aimed at building a new value profile has to take into account, at the greatest extent, both:

- the functional features of the subjected attributes, by choosing the most advantageous measures, according to the NVPGs, or at least avoiding the patterns viable to generate the biggest harm;
- the market-related evaluations regarding the attributes, by introducing absent and besides promising aspects, by boosting the features supplied at an unsatisfying performance level and by dedicating less effort to fulfill the customer requirements which outstrip the demand or however kindle minor attention.

Thus, the ultimate goal is the application of the NVPGs without infringing the fundamentals, although fuzzily formulated, of the original FAF.

The individuation of suitable attributes to be involved in the implementation of value-adding actions (Create and Raise) follows the product mapping finalized in the first Step of the IPPR procedure for class 3 of problems (Sect. 3.2.3.3). The choice of the focus advantages to be pursued within the accomplishment of the NVP task results the prior activity in addressing the switchover towards the projected product profile. At this stage, the IPPR user defines the basic actions characterizing the value transition, striving to identify, given the known technology, preliminary conceptual solutions and one or more market segments, capable to represent the first adopters of the enhanced business. The planned beneficial attainments are favorably accomplished without resorting to negatively impacting actions, especially those specified in Table 3.6.

In order not to build an unrealistic value changeover, the definition of the NVP architecture has to alternate the fine-tuning of a comprehensive list of actions associated with the correlated attributes and cycles aimed at delineating, though approximately, a physical idea. The latter can be properly supported by the tools suggested in the Step 3 of IPPR: further details about the subject are provided in Sect. 3.4.2.

Within the scope of the present Section, IPPR advices on the roadmap to be followed for achieving a potentially successful value profile:

(1) identify the main directions for product reengineering in compliance with the value-adding actions (Raise, Create) foreseen by NVPGs and FAF, highlighting whether the attained benefits are likely to be perceived by the whole market or some niche;

(2) generate a preliminary, although fuzzy, conceptual idea about how the previous selected actions (accompanied by the attributes) could be implemented, avoiding sophisticated or front-end technologies;

(3) check out whether the basic idea for product development could involve value reducing actions (Reduce, Eliminate) infringing the FAF or potentially revealing serious inconveniences according to NVPGs; reformulate the basic idea at step (2), if the disadvantages result excessively severe;

(4) write down each newly introduced action, updating the list of measures attained by the present NVP;

(5) progress towards a clearer physical solution that exploits the so far generated sample of actions, attempting to avoid those action potentially leading to the failure of the NVP initiative;

(6) check out whether the new product configuration gives rise to novel actions, regardless they are meant to enhance or reduce the customer satisfaction; update the list;

(7) verify whether the number of value-adding actions is consistently greater than those determining disadvantages and if the presence of not-compliant measures is marginal; in positive case, adopt the performed NVP, otherwise enrich the delivery of benefits restarting from point (1) or improve the product embodiment, overcoming the current shortcomings iterating from step (5).

3.4 Implementation of the "Ideal Solution to Physical Solution" Phase

The last step of IPPR is related to the identification of suitable physical solutions for the implementation of the new process, if the reengineering effort is within the scope of the classes 1 or 2, or the design of a novel product concept in case of the belonging to the 3rd group of business problems. The purpose of the Section is to identify the most suitable well-acknowledged tools coping with the indications about reengineering actions to be undertaken, as emerged from the previous step.

According to this objective, Sect. 3.4.1 suggests possible instruments to design the new industrial process while, Sect. 3.4.2 gives an overview of the tools that can support the conceptual design of a new product idea.

3.4.1 Guidelines for the Selection of the Process Redesign Tools

Here in the followings, selection criteria are provided in order to individuate the suitable instruments to finalize the reengineering task. The suggested tools aim at supporting the user in the redesign of the industrial process according to the actions emerged from the IPPR analyses.

As widely described in the previous Section, the directions which arise from the *Problem to Ideal solution* step to attain the improvement of a process may be briefly summarized as:

- enhance the Overall Value of the phase;
- trim the phase and, in such an event, assign the function to another process segment.

If the increment of the Overall Value is addressed at the growth of the arisen benefits, the strategy to be followed concerns either the enhancement of the phase performances or the exploitation of its capabilities to provide new customer requirements. In this situation the user has to investigate physical solutions aimed at enhancing the potentialities of the available technology or identify new ways for the implementation of the same function. Among the several contributions available in the literature, the *Classes 1.1* and *2* of the *76 Standard Solutions* belonging to the TRIZ body of knowledge [4], represents a viable design tool to attain the proposed objectives.

When a phase to be reengineered shows a high resources consumption, the objective of improving the Overall Value rate can be accomplished by identifying solutions which result more efficient. If the speed to introduce the product in the marketplace, or the timeliness of the goods delivery, represent the most critical issues, the design of new scheduling layouts aimed at minimizing the operational times can be supported by *Quick Response Manufacturing* [10]. In different circumstances, emerging undesired effects can represent a significant concern for the display of the business process. In this case the rethinking of the phases constituting the process bottleneck should be addresses towards a solution that conciliates the delivery of the attained benefits without provoking the manifested drawbacks. In order to pursue a more ideal solution, rather than a tradeoff between the extent of useful and harmful outputs, TRIZ tools are recommended to overcome the contradiction between the emerging pros and cons. In this perspective, the Algorithm for Inventive Problem Solving (ARIZ) [4] represents the most appropriate instrument to be implemented. Whereas the level of monetary expenditures represents the most evident cause of the process criticalities, the channeling of the resources has to be reorganized with the greatest priority. The *Lean Manufacturing* [11] provides a set of instruments capable to guide the designer in the identification of the resources which can be saved by the process, since they poorly impact the value for the customer. The *Class 2* of the *76 Standard Solution* provides suitable directions for reorganizing the flow of the resources with a particular focus on those resulting underused and on the process wastes. More specifically such tool guides towards the employment of underexploited resources for sustaining new additional industrial operations capable to provide further benefits for the product buyer.

Eventually, the results of the analysis of the business process may suggest to attain improvements by trimming phases showing very low performances. In addition, such strategy requires the assignment of the performed functions to other phases of the process. The task therefore necessitates the identification of substitution

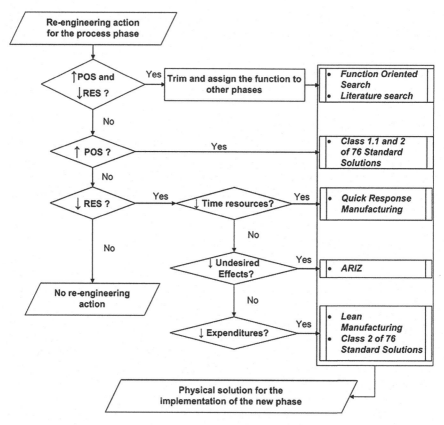

Fig. 3.13 Schema for the selection of the most suitable design tools according to the reengineering action to be implemented with reference to the results of the process analysis

technologies capable to satisfy the supplementary functional requirements without considerably impacting the resources consumption. To this end, a candidate support is represented by the *Function Oriented Search* approach [12], that allows to discover suitable technical solutions on the basis of the functional requirements to be fulfilled. An aid can be provided through the literature search: the consultation of scientific and technical databases, design catalogues and other sources of information has to be conducted with the aim of gathering sufficient knowledge for implementing a suitable technical solution. The proposed approaches involve the exploration of large amounts of data in order to extract the relevant information, thus the task can be eased by employing proper Knowledge Management systems.

Figure 3.13 depicts the selection path for the tools described so far, according to the kind of redesign action to be implemented for the phases to be reengineered, as arising from the observation of the PRAC and/or the low performance quadrant of the VAC diagram. In the latter case, with reference to the cells of the depicted flow chart highlighted in grey, the concurrent presence is excluded of sufficient delivered benefits and reduced resources consumption.

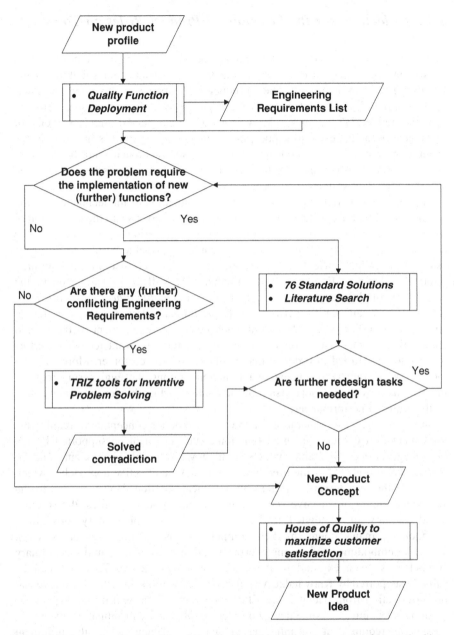

Fig. 3.14 Schema for the selection of the most suitable design tools to implement the hints emerging from the generation of an innovative product profile

3.4.2 Guidelines for the Selection of Product Redesign Tools

The output provided by IPPR with regards to the reengineering problems of class 3, consists in a set of new customer requirements to be translated into suitable product features. The identification of a new product idea capable to implement the innovative sample of product attributes at the required level of performances, refers to what the engineering literature acknowledges as a conceptual design task. Plenty of tools have been suggested to support the reengineering initiatives involved in the generation of new product concepts. Some valuable approaches are hereby suggested in order to orient the user towards the most effective tools according to the specific objectives of the design problem phase.

A preparative activity preceding the conceptualization of the technical solution is represented by the clarification of the design task. It consists in the translation of the customer requirements into well defined engineering characteristics. The detailed analysis of such features aims at identifying possible hurdles to fulfill the established product attributes. The output of this phase consists in a list summarizing the identified Engineering Requirements as well as a clear vision of the complexity of the design task. In such context, *Quality Function Deployment* (QFD) [3] is tailored to perform such preliminary design task. The model employed by QFD is capable to link each customer requirement to the related engineering characteristics, thus highlighting the most relevant technical performances to be fulfilled in order to maximally satisfy the customer. Moreover, the tool allows to remark the negative interactions among the engineering requirements, thus providing a clear vision of the conflicts which hinder the achievement of the required performances.

In some cases, the generation of the novel product concept may require the implementation of unprecedented functions; such task can be supported by the knowledge already available, which could however result insufficient. The *76 Standard Solution*s [4] or the literature search may be effective approaches which can guide the user towards the feasible technical solution. However, due to the variegated nature of the involved attributes within product profiles, the solution could require the knowledge dispersed across complementary disciplines (technology, management, market, computer science, human resources, etc.). Thus, the embodiment of an engineering solution may involve multidisciplinary competences, even external to the design team knowledge. This task can be suitably supported by Knowledge Management (KM) tools, in order to retrieve and use information from patents, scientific journals etc. even with limited resources.

In other circumstances the design problem is affected by incompatible extents of engineering requirements, meaning that a certain design choice allows the fulfillment of a given attribute, but results in disregarding other demands. Whereas one or more conflicts among the engineering characteristics come out, the design process requires to overcome these contradiction in order to generate a solution concept capable to conciliate the diverging requests. In such context, the tools provided by TRIZ to dig and solve technical contradictions represent an effectual aid to attain innovative ideas.

According to the above described criteria, Fig. 3.14 depicts a selection diagram supporting the identification of the most suitable tool according to the kind of design problem to be addressed. The flowchart terminates with the advised employment of the *House of Quality* [3] in order to set the relevant design parameters at such an extent to maximize the customer satisfaction. The task strengthens the definition of the physical solution implementing the innovative product profile.

References

1. Cavallucci D, Khomenko N (2007) From TRIZ to OTSM-TRIZ: addressing complexity challenges in inventive design. Int J Prod Dev 4(1/2):4–21
2. Becattini N, Cascini G, Petrali P, Pucciarini A (2011) Production processes modeling for identifying technology substitution opportunities. In: Proceedings of the TRIZ-future conference 2011, Dublin, 3–4 November 2011
3. Akao Y (1972) New product development and quality assurance: system of quality function deployment. Stand Qual Control 25(4):9–14
4. Altshuller GS (1984) Creavity as an exact science. In: Williams A (trans) The theory of solution of inventive problems. Gordon and Breach Science Publishers, New York
5. CQM (1993) A bose development teams experience with Kano mapping. Cent Qual Manag J. Obtained through the internet: http://www.walden-family.com/public/cqm-journal/2-4-Whole-Issue.pdf. Accessed 05 Dec 2011
6. Tontini G (2003) Deployment of customer needs in the QFD using a modified Kano model. J Acad Bus Econ 2(1):103–113
7. Mittal V, Ross WT, Baldasare PM (1998) The asymmetric impact of negative and positive attribute-level performance on overall satisfaction and repurchase intentions. J Mark 33(4):271–277
8. Borgianni Y, Cardillo A, Cascini G, Rotini F (2011) Systematizing new value proposition through a TRIZ-based classification of functional features. Procedia Eng 9:103–118
9. Borgianni, Y, Cardillo A, Cascini G, Rotini F (2011) Design of innovative product profiles: anticipatory estimation of success potential. In: Proceedings of the 18th international conference on engineering design, Copenaghen, 15–18 August 2011
10. Suri R (1998) Quick response manufacturing: a companywide approach to reducing lead times. Productivity Press, Portland
11. Womack J, Jones D (1996) Lean thinking. Simon and Schuster, New York
12. Litvin SS (2004) New TRIZ-based tool—function-oriented search. In: Proceedings of the 4th TRIZ future conference, Florence, 2–5 November 2004

Chapter 4
Application of IPPR to the Reengineering Problems of Class 1

4.1 Introduction: The Italian Industry of Woody Bio-Fuel

In this Chapter the application of the methodology is presented to the woody pellet production process. This sector presents high business opportunities in Italy since the market demand of such kind of energy sources is grown dramatically in the last five years. Besides, the industrial processes treating widely available (but not optimal) resources are still under development. Their poor performance does not allow the attainment of an output complying with the requirements imposed by regulations and standards currently in force, if not resorting to heavy expenditures. Consequently, the business process is not able to fully exploit the available biomass resources, giving rise to the impossibility to meet the unsatisfied market demand of woody fuels.

In such context, the applicability of the developed approach to industrial problems originated by this kind of under capacities has been tested.

The content of the Chapter is structured in three main parts. Section 4.2 reports an overview of the faced problem, delving into its main critical aspects. The Sect. 4.3 presents the application of the IPPR methodology with the aim of showing how the suggested tools are employed. Eventually, Sect. 4.4 provides a brief discussion on the consistency of the obtained results.

4.2 General Overview of the Business Process

The solid bio-fuel coming from the sustainable exploitation of forest resources represents a not negligible complementary source of energy to oil and its derivatives. In the last years the market demand of such a resource is dramatically grown, with a particular reference to Italy, resulting in a business opportunity for several rural areas: one of these is the Tosco-Emiliano Apennine, a mountainous territory in the north-central part of the country.

F. Rotini et al., *Re-engineering of Products and Processes*,
Springer Series in Advanced Manufacturing, DOI: 10.1007/978-1-4471-4017-7_4,
© Springer-Verlag London 2012

Two different kinds of bio-fuel are obtained by the sustainable exploitation of the forest resources:

- *wood chips*: pieces of wood having overall dimensions of 25 × 30 × 20 mm, maximum moisture content of 20% in weight, average market price 70 €/ton;
- *pellets*: cylinders of pressed sawdust having a diameter of 6 or 8 mm, height of 35 mm, moisture content of 10% in weight, average market price 180 €/ton.

Both the products have mandatory characteristics prescribed by specific standards [1]. More precisely, the pellet must fulfill the following main requirements:

- Lower Heating Value: >18 MJ/kg.
- Shape and size: cylinder with dimensions of φ6 × 35 mm.
- Sufficient mechanical resistance in order to avoid the breaking during transportation and feeding of the burning system.
- Good capability to keep a constant energetic content.

Table 4.1 shows an example of local exploitation of biomass resources, referred to a delimited area located in the Tosco-Emiliano Apennine. In this region the amount of biomass obtained by the sustainable exploitation of forests may constitute an energy source capable to satisfy the needs of about 6600 housing units making them almost independent from the oil derivatives. Altogether, the resources available for the manufacturing of bio-fuels are essentially sawdust and waste obtained by the maintenance operations of the forests and the urban green. The sawdust comes from wood industry and is characterized by a low content of the moisture. The waste is supplied in form of pieces of tree, which usually own high moisture content.

A preliminary analysis of the business process showed that the wood waste is mainly used to manufacture chips, while pellets are basically obtained through the transformation of sawdust. As shown in Table 4.1, the yearly availability of the wood coming from sawmills is smaller than the amount coming from forest and urban management. Actually the business concerning the production of woody fuels is able to satisfy the market request of wood chips, while a big deal of the demand of pellets remains unmet.

From a technological point of view, the pellet manufacturing process employs knowledge coming from industrial fields which show severe differences with reference to the wood manufacturing sector. As detailed further on, the process is constituted by three main activities, consisting in the grinding, dewatering and pressing of the wood.

The grinding is actually performed through hammer mills which are meant to crush brittle dry materials in fine particles. Unfortunately, when such systems are used to crush the wet wood, they frequently clog up due to the formation of a mush that interrupts the flow of the material inside the machine.

The dewatering phase is performed by a thermal dehumidification employing "traditional" ovens which burn oil, methane or a part of the raw material. Due to the high moisture content that has to be removed from the wood, this phase involves a high energy consumption. Moreover, the temperature that the biomass

Table 4.1 Woody biomass resources available in the Tosco-Emiliano Apennine (tons/year)

Origin	Moisture content (in weight) (%)	Estimated availability	Estimated availability after 10 years
Wood coming from industry processes	10	5000	6000
Wood coming from forest management	35–50	25000	50000
Wood coming from urban green management	45–50	2000	10000

reaches inside the oven is a critical process parameter. If the temperature exceeds a certain limit, the dewatering phase can result in the reduction of the energetic content of the wood, because of the detaching of volatile substances such as alcohols. On the contrary, if the temperature inside the oven is too low, the dehumidification process is not able to reduce the moisture content of the wood at the required extent.

Eventually, the pressing of the sawdust is performed by machines developed within the animal feed industry, whose input raw material shows properties that consistently differ from the characteristics of the wood (mushiness, mechanical features).

As a result, these imported technologies have demonstrated low performances and efficiency.

The exposed considerations highlight that the satisfaction of the pellet market demand merely depends on the capability of the process to handle the wood waste. With reference to such issue, relevant under capacities emerge, which hinder the exploitation of the green wood as primary source of raw material for the pellet production. Since the business limitations remarkably depend on the capability of the process to provide the desired output rather than on the lack of competitiveness, the considered industrial problem can be referred to those falling into the class 1.

4.3 Application of IPPR

The application of IPPR methodology has been performed according to the roadmap developed to address the problems belonging to the class 1. For the sake of clarity the followed procedure has been reported in the Table 4.2.

In the following Sections the application of IPPR is described, clarifying the usage of the tools presented in the previous Chapter.

4.3.1 Process to Problem

As foreseen by the methodological flow, the first activity of IPPR involves the modeling of the industrial process, schematizing the constituent phases and the

Table 4.2 IPPR methodology tailored on the pellet manufacturing reengineering problem

Phase	IPPR activity	Tools
Step 1		
Process to problem	Process modelling	• Multi-domain modeling technique
	Product information elicitation	• Correlation coefficients
	Product modeling	• Relevance scale
Step 2		
Problem to ideal solution	Identification of what should be changed in the process	• Phase Overall Satisfaction metric
		• Resources consumption metric
		• Value indexes
		• PRAC Diagram
Step 3		
Ideal solution to physical solution	Finding physical solutions for new process implementation	• Guidelines to select process redesign tools

provided outputs. The Multi-domain modeling techniques is conveniently adopted to carry out the functional representation of the industrial process, which summarizes all the relevant information pertaining the displayed phases. Moreover, the analysis of the transformations operated by each phase is a fundamental issue for the subsequent tasks, since it highlights how the activities participate to the generation of value in terms of customer requirements. The impact of the analyzed industrial activities on the determination of the product attributes is quantified through the Correlation coefficients. Eventually, the model of the product reveals by means of the Relevance scale the importance of each customer requirement in the value building.

4.3.1.1 Process Modeling

The accomplishment of the process schematization task through the multi-domain model requires the identification of the phases and of the involved resources. The segmentation of the process into the constituent activities can be easily performed by mapping the transformations of the relevant properties characterizing the raw material and its intermediate states.

As recalled in the previous Chapter, the *Element Name Value* model provides an aid in fulfilling the task. The raw material processed during the manufacturing of pellets is constituted by wood whose properties are modified as described in the following sequence:

(1) the wood pieces undergo a reduction of the average size from 100 to 30 mm and, at the same time, a mild decrease of the moisture content from 50 to 45% in weight;

(2) the impurities inside the raw material are removed until the output reaches a degree of purity equal to the 99%;

(3) the resulting material is subjected to a further drastic reduction of the water content from 45 to 15%;

(4) the wood size is reduced to that of the sawdust (2–5 mm) and the residual moisture is adjusted to the level required for the final product (10%);

(5) the resulting sawdust is transformed in small cylinders constituting pellets, which have a diameter of about 6 mm and a length of 35 mm;

(6) eventually, the temperature is reduced from 80 to 20°C so that the pellets can be packaged in bags containing approximately 15 kg of bio-fuel.

Therefore, the analysis of the transformations clearly highlights six main phases into which the pellet manufacturing process can be segmented:

- the first trituration
- the subsequent purification
- the drastic moisture reduction through dewatering
- the second trituration
- the pressing of the sawdust namely pelletizing
- finally, the cooling and packaging of the pellet

The identified phases and the properties of the raw material that are modified, can be summarized as shown in Table 4.3, in coherence with the recalled concept of the Element Name Value.

Once the phases that form the process have been formalized, it is possible to identify the other resources involved. Since the under capacities affecting the process do not depend on the productivity, nor the time to market is a key issue, the duration of the phases in completing the assigned activity does not constitute a relevant parameter for the scope of the process analysis, thus it has been neglected. Moreover, the undesired flows have not been represented due to the absence of severe process inconveniences. Consequently, the following flows have been collected for each phase:

- Energy
- Occupied space
- Materials
- Involved human resources
- Involved technologies and know how
- Information
- Control parameters of the phases.

Table 4.4 summarizes the power consumption, the employed human resources, the required space for the equipment and machinery and other employed materials that cannot be ascribed to any of the previous categories. The quantities have been indicated according to the processing of wood giving rise to the production of 1000 kg of pellet. It is worth to notice that the dewatering phase requires a high energy consumption in order to reduce the moisture content of wood chips from 45

Table 4.3 Phase, processed flow, modified parameters of the flow, input and output values

Phase	Element	Name	Input value	Output value
A1—Trituration	Wood	Size	100 mm	30 mm
		Moisture content	50%	45%
A2—Purification	Wood	Purity	80%	99%
A3—Dewatering	Wood	Moisture content	45%	15%
A4—Second trituration	Wood	Size	30 mm	2−5 mm
		Moisture content	15%	10%
A5—Pelletizing	Sawdust	Size	2−5 mm	–
	Pellet	Shape	Undetermined	Cylinder ϕ 6 × 35 mm
A6—Cooling and packaging	Pellet	Delivering status	Untied pellets	Pellets: bags of 15 kg
		Temperature	80°C	20°

to 15% in weight. Furthermore, the pelletizing is accounted to a considerable human involvement in terms of labor, experience and know how, while the machines to perform the dewatering show the largest size. Finally, the packaging requires the purchasing of bags.

The involved technologies and the parameters governing the display of the process segments, are summarized in Table 4.5. The chipper is a cutting machine allowing the shredding of big pieces of wood into fragments. It is constituted by a rotating disc which carries cutting knifes. The purification of the wood is performed through a separator which extracts the ferrous impurities and sieves stones, soil and other residual that can compromise the quality of the bio-fuel. The dewatering phase is implemented by the usage of an oil or methane kiln, whose temperature is kept between 150 and 300°C. The reduction of wood chips into sawdust is performed by means of a mill constituted by several rotating hammers that fulfill the grinding of the chips into fine particles. A sieve guarantees that the processed material is conveyed to the following machinery only when the sawdust has reached the required size. Subsequently, the pelletizing machine presses the sawdust through calibrated holes obtained on a die and cuts the extruded material at the right length, so to get the pellets, whose surface gets quite waterproof ad a result of the operation. Eventually, after the cooling of the material carried out to avoid the melting of the bags, the packaging is performed through a machine which weighs out a quantity of pellets equivalent to 15 kg.

The collection of the above data has allowed the construction of the process model employed to manufacture the pellets according to the formalism explained in the previous Chapter. Figure 4.1 depicts the model of the A1 and A2 phases, Fig. 4.2 shows the phases A3 and A4 while Fig. 4.3 presents the model of the phases A5 and A6. It is worth to notice that the multi-domain model is capable to provide an exhaustive overview of the industrial process, allowing to focus on the relevant information within the scope of IPPR.

Table 4.4 Summary of the resources which the process uses for the manufacturing of the pellet (beyond the wood)

Phase	Energy (kW/ton)	Labour (# employees)	Space (m^2)	Materials
A1	15	2	6	–
A2	10	2	12	–
A3	350	6	21	Natural air
A4	15	2	6	–
A5	55	10	6	Natural air
A6	7.5	2	9	Bags

Table 4.5 Technologies adopted for the implementation of the process phases and relevant control parameters governing their working

Phase	Machinery	Control parameter
A1	Chipper	Cutting speed
A2	Sieving and magnetic separator	Size of the sieve
		Magnitude of the magnetic field
A3	Oven	Inlet temperature
A4	Hammer mill	Cutting speed
		Size of the sieve
A5	Pelletizing machine	Size of the calibrated holes
		Cutting length
A6	Packaging machine	Cooling time
		Packaging speed

4.3.1.2   Product Information Elicitation

Within the list of the product features intended to create value, a relevant set of attributes emerges by taking into consideration the requirements prescribed by regulations and standards. As mentioned in Sect. 4.2, the mandatory characteristics imposed in this industrial sector regard:

- a suitable energy content, in terms of *Lower Heating Value, LHV* (CR1)
- *Shape and dimensions* (CR2)
- *Mechanical resistance* (CR3)
- *Capability to preserve the heating characteristics* (CR4).

The identification of a more comprehensive record of factors that contribute to generate satisfaction is carried out by exploiting the model of the process and thinking over, in the buyer perspective, the motivations underlying the occurred transformations.

As an example on how to follow this survey approach, let's consider the *Cooling and Packaging* phase. According to the parameters summarized in Table 4.2, the phase changes the "*delivering status*" of pellets from "*untied*" to "*in bags of about 15 kg*". The objective is to discover the motivations behind the changing of this property from the viewpoint of the exigencies to be fulfilled or

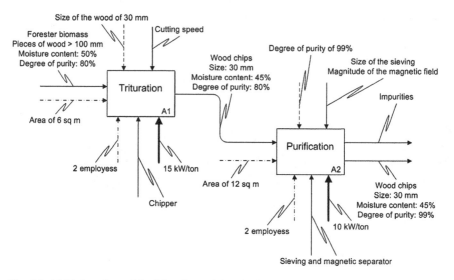

Fig. 4.1 Multi-domain model of the phases A1 and A2

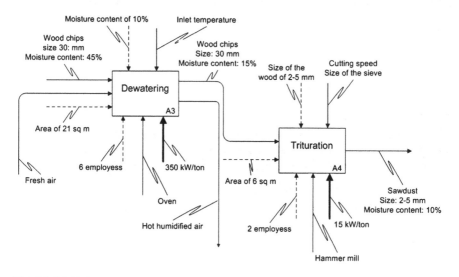

Fig. 4.2 Multi-domain model of the phases A3 and A4

benefits obtained by the customer. The formulation of questions, like the one that follows which concerns the specific modified parameter, is an useful reasoning tool for elucidating the value aspects the process is conceived for:

Within the perspective of value delivery which reason or scope motivates the transformation of the "delivering status" of pellets "untied" into "in bags of about 15 kg" along the analyzed phase?

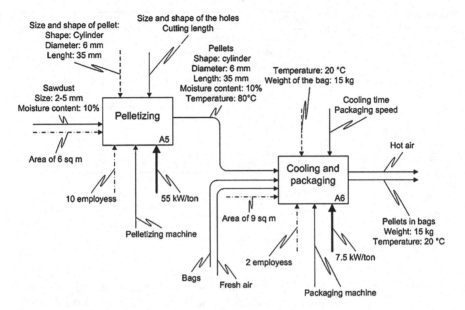

Fig. 4.3 Multi-domain model of the phases A5 and A6

The answer to this question brings to the identification of the benefits for the end-user which justify the existence of the Cooling and packaging phase. In this case the supplying of the pellets in bags of 15 kg allows an easy storage within the domestic stoves or inside the other burning systems currently available. The interest for such characteristic is confirmed also by the marketplace where the pellets are prevalently sold in bags rather than loose. Hence, the considerations performed so far suggest the "availability of the pellet in bags" as further requirement to be introduced in the attributes sample.

Beyond the just recalled requirement, the analysis of the modified parameters summarized in Table 4.3 has not highlighted additional features than those dictated by the regulations. Thus, it can be concluded that the customer requirements to be satisfied are those reported in Table 4.6.

Once the relevant product features have been individuated, it is possible to map the contribution of each phase in determining each customer requirement. In order to accomplish this task, the assessment of the relationships among the parameters modified along the process and the product features facilitates the determination of the estimated Correlation coefficients. A couple of examples is provided, concerning CR1 and CR4, about the way of computing the contribution of the phases in the fulfillment of the requirements throughout the consideration of the engineering characteristics that are accomplished. Table 4.7 monitors which technical features impact the attainment of each customer requirement. Subsequently, the involvement of each phase in obtaining each customer requirement is summarized in Table 4.8.

Attribute
CR1—Lower heating value—LHV
CR2—Shape and dimensions
CR3—Mechanical resistance
CR4—Capability to preserve the heating characteristics
CR5—Availability in bags

Table 4.6 Customer requirements to be satisfied for the pellet

Modified parameters	CR1	CR2	CR3	CR4	CR5
Size		X			
Moisture content	X				
Purity			X	X	
Shape		X	X	X	X
Delivering status					X

Table 4.7 Relationships among modified properties of the main flow of raw material and the customer requirements

Table 4.7 plainly shows that the unique technical parameter impacting the energy content of the pellet is represented by the moisture included in the wood. It can be thus stated that the phases contributing in the attainment of CR1 are those which modify the moisture content of the processed wood, hence A1—Trituration, A3—Dewatering and A4—Second trituration. The desired level of water removal is obtained by the partial contributions depicted in Table 4.3, hence:

- Moisture reduction operated by the phase A1 (Δ_{A1}): 5%
- Moisture reduction operated by the phase A3 (Δ_{A3}): 30%
- Moisture reduction operated by the phase A4 (Δ_{A4}): 5%

Therefore, the values of the correlation coefficients pertaining the CR1 and the phases A1, A3 and A4 are calculated as in the followings:

$$k_{11} = \Delta_{A1}/(\Delta_{A1} + \Delta_{A3} + \Delta_{A4}) = 5/40 = 0.12 \approx 0.1$$
$$k_{13} = \Delta_{A3}/(\Delta_{A1} + \Delta_{A3} + \Delta_{A4}) = 30/40 = 0.75 \approx 0.8$$
$$k_{14} = \Delta_{A4}/(\Delta_{A1} + \Delta_{A3} + \Delta_{A4}) = 5/40 = 0.12 \approx 0.1$$

Furthermore, if we consider the CR4—Capability to preserve the heating characteristics, such a feature is guaranteed through a high degree of the wood purity and by waterproof characteristics of the pellet surface. As suggested in Chap. 3 for the cases ascribable to this situation, the adopted criterion for the calculation of the correlation coefficients is based on a weighted sum of the partial contributions offered by each property in determining the analyzed CR. Specifically, the capability to preserve the heating characteristics are influenced much more consistently by the waterproofing of the pellet surface rather than on the wood purity. Consequently the former engineering characteristic holds a greater relevance (80% as estimated by sector experts) than the latter in the attainment of the treated customer requirement. The required purity of the wood is achieved

Table 4.8 Role played by each phase in generating each customer requirement

Phase	CR1	CR2	CR3	CR4	CR5
A1—Trituration	X	X			
A2—Purification			X	X	
A3—Dewatering	X				
A4—Second trituration	X	X			
A5—Pelletizing		X	X	X	X
A6—Cooling and packaging					X

through the Purification (A2); conversely the requested waterproof qualities are attained through the Pelletizing (A5). It can be concluded that the CR4 is determined as a result of the concomitance of the two cited phases. According to the performed considerations, the extent of the contributions of the Purification and of the Pelletizing, can be calculated as follows:

- Improvement of the degree of purity performed by A2 (Δ_{A2}): 100%
- Improvement of the degree of purity performed by A5 (Δ_{A5}): 0%
- Attainment of the desired level of waterproof operated through A2 (Δ_{A2}): 0%
- Attainment of the desired level of waterproof operated through A5 (Δ_{A5}): 100%
- Importance of the degree of purity in determining the CR4 (W_1): 0.2
- Importance of the waterproof properties in generating the CR4 (W_2): 0.8

$$k_{42} = \Delta_{A2}/(\Delta_{A2} + \Delta_{A5}) \times W_1 = 0.2$$
$$k_{45} = \Delta_{A2}/(\Delta_{A2} + \Delta_{A2}) \times W_2 = 0.8$$

The calculation of the further correlation coefficients according to the suggested criteria have led to the values summarized in Table 4.9.

4.3.1.3 Product Modeling

The present activity concerns the determination of the importance that each customer requirement owns in impacting the value perceived by the end-user. According to the indications suggested in Chap. 3, this task is carried out by assigning a score to each product feature, that is expressed through a natural number included in the interval 1–5 of a Likert scale.

Since the heating capability is the characteristic of the pellet which is mostly appreciated by the customer during the use of such bio-fuel, representing the main provided benefit, its degree of importance can be evaluated with the maximum score. The size, the mechanical resistance and the capability to preserve the heating power can be accounted as requirements aimed at prevalently meeting the functional exigencies of the burning systems. Although these features strongly contribute to the prevention of safety problems, their offering does not impact the perception of the customer at the maximum extent. Therefore, notwithstanding the

Table 4.9 Correlation coefficients quantifying the relationships phase versus customer requirements

Phase	CR1	CR2	CR3	CR4	CR5
A1	0,1	0,2	0	0	0
A2	0	0	0,1	0,2	0
A3	0,8	0	0	0	0
A4	0,1	0,1	0	0	0
A5	0	0,7	0,9	0,8	0,1
A6	0	0	0	0	0,9

Fig. 4.4 Relevance coefficients expressing the importance of each customer requirement in determining the customer perceived value

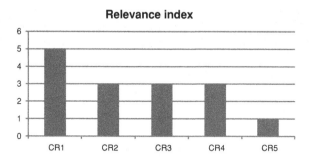

importance of the role played by these characteristics, they are assigned a relevance index inferior than that of the Lower Heating Value. Finally, the delivery of pellets in bags is a feature which facilitates the usage of the product, by easing its storage, transportation and handling. It can be considered as an additional feature whose presence is appreciated by the customer, but, however, its relevance in determining the perceived value is considerably more limited with respect to the other characteristics.

According to the above performed considerations, the assigned relevance indexes are summarized in Fig. 4.4.

4.3.2 Problem to Ideal Solution

The tools suggested by PVA are hereinafter applied in order to identify the main process under capacities which hinder the attainment of the desired output of the process, according to the resources available for the pellet manufacturing.

4.3.2.1 Phase Overall Satisfaction

The calculation of the *Phase Overall Satisfaction* (POS), representing the extent that each activity has in determining the customer contentment, is determined through the following formula:

$$POS_j = \sum_i k_{ij} \times R_i.$$

Phase	POS$_j$	POS$_j$(%)
A1—Trituration	1,2	8,2
A2—Purification	0,9	6,0
A3—Dewatering	3,8	25,0
A4—Trituration	0,9	6,2
A5—Pelletizing	7,3	48,7
A6—Cooling and packaging	0,9	5,3

Table 4.10 POS of the j-th phase

Table 4.10 summarizes the obtained POS$_j$ of each process phase (expressed in a non-dimensional form and as a percentage).

The POS coefficient highlight the central role of A1, A3 and A5 in determining the customer contentment. Besides, as demonstrated by the k$_{ij}$ coefficients and the relevance indexes R$_i$, they are fundamental in the attainment of the most relevant product features.

4.3.2.2 Resources Consumption

The calculation of the *Resources consumption* (RES) index, has to be performed through the usage of the expression introduced in Chap. 2:

$$RES_j = c \times C_j + t \times T_j + h \times HF_j.$$

However, as recalled in the previous section, the time to market does not represent a critical factor for the analyzed business process; moreover, the manufacturing of pellets from wood waste does not show remarkably lower speed than the traditional production based on sawdust. Hence, the duration of the production phases is not considered relevant for the examined value creation process, as well as the undesired effects emerging by the sequence of the process segments (e.g. noise, vibrations, difficulties related to the maintenance, etc.), that are not actually displayed. Therefore, the global resources estimation has neglected operating times and drawbacks, focusing the attention just on the monetary expenditures. In other terms, the coefficients, t and h are null in the specific case study.

The extent of the channeled resources has been assessed by assuming a reference production of 1 ton of pellet. More specifically, the analysis has included the expenditures related to the labor, the space occupied by the equipment, the consumed energy and materials, subsequently evaluated through a monetary metric.

The inventory costs related to the acquisition of the necessary quantity of wood have been neglected since such kind of biomass is still considered alike waste and currently it does not possess any economic value. The energy costs have been calculated with reference to the consumption of each phase and the current price of the electric/thermal power. The operating expenses related to the labor have been calculated through the accounted involvement of the personnel and the hourly cost of the employed workers. The inventory costs involved for the space occupied by the plants have been calculated dividing the monthly amount of real estate expenditures for the industrial site by the potential production of the plant in the same period. Then, such expenditures have been split to calculate the amount accounted to each process step taking into consideration the ratio of the space occupied by the machinery utilized to perform the phases. Finally, the inventory costs of the consumed materials have been assessed dividing the annual expenditures to purchase these resources by the number of pellet batches potentially produced in a year.

Thus, the expenditure values have been calculated through the following rules:

- *Energy expenditures = phase required energy × power cost;*
- *Labor expenditures = phase employed labor hours in a year × hourly labor cost index/number of pellet batches produced in a year.*
- *Space expenditures = (ratio of space occupation for the phase machinery) × (monthly real estate expenditure)/(number of pellet batches potentially produced in a month).*
- *Material expenditures = costs to purchase the needed materials for a ton of manufactured pellet.*

Some examples are herein proposed in order to clarify the assessment procedure of these expenditures. The considered values for the resources consumption are those summarized in Table 4.4.

With reference to the A1 phase, the cost to purchase 1 kW of electric power is about 0.2 €/kW, hence the expenditures for the trituration of 1 ton of wood are:

$$Energy\ expenditures\ (A1) = 0.2 \times 1.5 \approx 3\,€/\text{ton}$$

On the other hand, if the cost of 1 kW of thermal power needed by the oven is 0.11 €, the energy expenditures for the Dewatering phase (A3) is evaluated in:

$$Energy\ expenditures\ (A3) = 0.11 \times 350 \approx 38\,€/\text{ton}$$

Moreover, if the hourly labor cost index is 17.5 €/h, each worker dedicates his/her full time to this operation (1516 h expected in a year), the yearly production can be estimated in 27000 tons, such kind of expenditures required by the Pelletizing phase can be evaluated in:

$$Energy\ expenditures\ (A2) = 2 \times 1516 \times 17.5/27000 \approx 2\,€/\text{ton}$$

Let's now take into account the expenditures involved in the Purification to "acquire" the needed space where the phase is performed. The number of pellet batches produced in a month can be easily obtained dividing the yearly availability of wood resources by the number of labor months included in a year:

$$Number\ of\ pellet\ batches\ potentially\ produced\ in\ a\ month = 27000/11$$
$$= 2454\,\text{ton/month}$$

Furthermore, the ratio of space occupation (as emerging from Table 4.3 by considering the room needed for the phase and that required for the whole process) for the employed purification machinery is:

$$Ratio\ of\ space\ occupation\ for\ the\ phase\ machinery\ (A2)\ =\ 12/60\ =\ 0.2$$

If the monthly real estate cost is quantified in 25 k€, the space expenditures for the Purification phase can be evaluated in:

$$Space\ expenditures\ (A2)\ =\ 0.2\ \times\ 25000/(2454)\ \approx\ 2\,€/ton$$

Eventually, considering the Packaging phase, if the price of a bag is 0.10 € and remembering that each bag can contain 15 kg of wood pellets, the expenditures of material are:

$$Material\ expenditures\ (A6)\ =\ 0.10\ \times\ 1000/15\ =\ 6,7\,€/ton$$

By following the explained criteria, the RES index of each phase has been evaluated. The obtained values are reported in the Table 4.11 (in €/ton and in percentage in the last column).

The evaluation of the resources consumption shows that, among the six phases of the process, the Dewatering of the wood and the Pelletizing are the most expensive with a massive impact of the former on the whole efficiency of the process. Furthermore, the overall cost to manufacture 1 ton of pellet exceeds half the price in the marketplace, giving rise to low margin of revenues for the business process.

4.3.2.3 Overall Value and PRAC Diagram

The *Overall Value* index of the process phases has been calculated as the ratio between the POS and the RES coefficients, both expressed as percentages. The resulting values are shown in the Table 4.12. Moreover, the *POS versus RES Assessment Chart* (PRAC), shown in Fig. 4.5, has been built with the aim of providing a clear representation of the business process analysis.

The conjoint analysis of the OV index and the PRAC leads to several directions of investigation devoted to the improvement of the manufacturing process.

As shown in Table 4.12, the dewatering (A3) owns the smallest OV index, thus it represents the major bottleneck of the business process. Such outcome emerges as a result of the substantial amount of resources dedicated to the working of the phase. Furthermore, the rank of the OV indexes shows that the second critical phase is the trituration (A4), which provides a poor contribution in determining the customer satisfaction and presents a limited consumption of resources. According to Fig. 4.5, the purification (A2), the packaging (A6) and first the trituration (A1) show remarkable analogies with the discussed (A4). Eventually, the pelletization (A5) is deemed as a phase which considerably contributes to the customer satisfaction.

Table 4.11 Resources consumption index of each phase

Phase	Energy	Labour	Space	Material	RES$_j$	RES$_j$(%)
A1—Trituration	3	2	1	–	6	6.1
A2—Purification	2	2	2	–	6	6.1
A3—Dewatering	38	5.9	3.5	–	47.4	47.9
A4—Second trituration	3	2	1	–	6	6.1
A5—Pelletizing	11	9.8	1	–	21.8	22
A6—Cooling and packaging	1.5	2	1.5	6.7	11.7	11.8

Energy, labour and space resources are expressed in €/ton

Table 4.12 Overall Value index of each phase

Phase	OV$_j$	OV$_j$(%)
A1—Trituration	0.38	34.3
A2—Purification	0.15	13.4
A3—Dewatering	0.06	5.7
A4—Trituration	0.12	10.4
A5—Pelletizing	0.33	30
A6—Cooling and packaging	0.07	6.1

The whole analysis suggests that primary fundamental action to be pursued consists in the improvement of the dewatering phases by developing more efficient technologies for wood pellet production. Moreover, given the characteristics of the (A4), strategies should be evaluated in order to migrate its function to other phases or integrate the delivery of additional benefits. Such indications pertain also the (A1), (A2) and (A6), although the reengineering efforts for such phases are not of overriding importance. Eventually, although the pelletization holds a high OV coefficient, the performed analysis reveals a not negligible resources consumption for this phase (Table 4.11), thus technologies enhancement aimed at further increasing its efficiency would be welcome in each case.

4.3.3 Ideal Solution to Physical Solution

The value-oriented process analysis has further shed light on the consistent limitations concerning the production of pellet starting from wood waste, due to the experimental technologies. The tools and machinery employed so far, with a particular emphasis on the dewatering process, result scarcely efficient to treat biomass with a high moisture content. In order to obtain pellet with a satisfying energetic yield, the moisture content initially present in the green biomass (approximately 50% in weight) must be drastically reduced. The technologies based on thermal dewatering use rotating or fluid bed furnaces that are fed by methane, oils, or a part of the raw biomass. This involves high fuel consumption, due to the meaningful

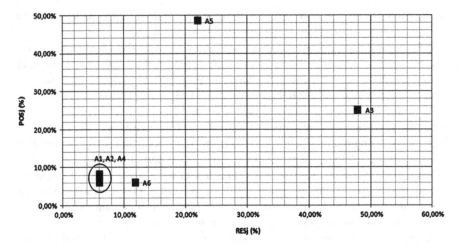

Fig. 4.5 PRAC of the pellet manufacturing process

amount of water that should be extracted. As experiences illustrate, a critical factor for the efficiency of the dehydration is the size of the treated material. Indeed, the dewatering phase could be strongly enhanced if the size of the biomass at the inlet of the kiln could be reduced, but, unfortunately the current systems for wood grinding are not able to treat wet biomass. On the other hand, the technologies currently positioned downstream of the dewatering that impact the structure of the biomass (i.e. the trituration and the pelletizing) do not show any opportunity to support the drying, so the bottleneck of the process can hardly be substituted.

According to outcomes of the analysis and the preceding considerations, the most beneficial directions to be followed in order to make the process convenient would regard the adoption of technologies capable to:

- dewater the chips, reducing considerably their moisture from 50% to very low values, by employing a smaller amount of energy than that required by the present systems;
- triturate the wood chips into finer particles during other process segments, favorably along a reengineered dewatering phase.

These exigencies constitutes well defined functional requirements that the new process must satisfy.

The definition of the projected process improvements has led to the formulation of two well specified technical problems, whereas the first holds primary importance:

(1) How is possible to reduce the resources consumption of the dewatering phase?
(2) Is it possible to integrate elsewhere the functions performed within the trituration? Can the redesign of the dewatering operations include the milling functions?

Generally speaking, such a design task is related to the identification of technical solutions capable to efficiently fulfill the expected phase performances, i.e. the trituration of the wood chips up to the sizes required for the pelletization as well as a suitable extraction of the moisture from the wood.

With regards to the tools selection criteria exposed in the Chap. 3, the prior redesign problem falls into the category of identifying technical solutions aimed at minimizing the process expenditures. Thus, the principles of *Class 2* with regards to the *76 Standard Solutions* of TRIZ, result as candidate techniques to support this task.

According to the exploitation of the mentioned tools, the research of working principles has been focused on the identification of alternative physical solutions for the dewatering phase. Specifically, high speed mechanical energy has resulted in a powerful resource to separate water from wood particles during the milling process. The individuation of such solution has brought about particular interest due to the possibility to integrate (at least partially) the trituration and the dewatering. Moreover, if ultrasonic waves are generated by means of high speed shocks, they can further contribute to the moisture reduction.

A specific patent search to validate such a conceptual solution has produced the individuation of three patents [2–4] adopting the same physical principle to pulverize and dry several kinds of raw materials. At least one of these patents has been converted into a real product [5]: a rotor equipped with chains or knifes operates the trituration of the material, by shooting the particles towards the walls of the machine. The impact transforms the kinetic energy of the particle into vibration energy, thus the particles and the water vibrates: this allows the separation of the different materials. According to the datasheet supplied by the producer, such a system is able to reduce the moisture content of the wood from 60 to 10% and the particle size up to 1 mm. The most relevant property of this technology is a very limited energy consumption, about three times less than a traditional heat based dehumidification.

Unfortunately such a technology is not suitable to be used during forestry operations since it has dimensions and weight that do not allow transportation and management in the forest areas. With the aim of overcoming this limitation, a new mechanical system which implements the same physical principle adopted in [5] through a different architecture so to avoid patent infringement, has been designed and developed by the authors and other colleagues [6]. Such a system has a size that allows its transportation and installation in the woody areas where forestry operations take place. This technology is less expensive than the traditional one in terms of both investments and maintenance costs. Tests also revealed that the biomass can reach the required moisture content for pellet production after very few milling/dewatering cycles.

4.4 Discussion of the Outcomes

The present Section discusses the reliability of the outcomes of the process analysis performed within IPPR, regardless the effectiveness of the extrapolated physical solution, for which further verifications are still required.

The analysis of the scientific and technical literature in the field of renewable energy confirms that the drying of the woody biomass is a critical phase in the production process of pellet starting from green wood.

In [7, 8] it is clearly explained that the drying process based on thermal heating has a not negligible impact on both quality and production costs of wood pellet and new drying systems should be developed in order to make the pellet manufacturing process more efficient in terms of energy consumption and product characteristics delivered to the end-user.

In [9] it is claimed that in wood manufacturing industry, drying is considered the most relevant matter determining problems in process controllability and high energy expenditures. Several studies have been carried out and several technologies have been introduced to improve this phase in wood industry, as summarized in [10], showing however the absence of a dominant design or standard.

Therefore, the aforementioned researches widely confirm the results obtained through the application of the proposed methodology.

References

1. Standard UNI 11263 (2007) Ente Nazionale Italiano di Unificazione
2. Sand et al. (1998) Device and method for comminution. US Patent 5,839,671, Spectrasonic Disintegration Equipment Corp
3. Sand J, Martin JE, Clarke-Ames JJ (2000) Device and method for comminution. US Patent 6,024,307
4. Hamm RL, Hamm GL (2000) Pulverizer. US Patent 6,039,277
5. First American Scientific Corporation (2011) Innovators in green fuel preparation. Obtained through the internet: http://www.fasc.net/. Accessed 30 Nov 2011
6. Tonarelli A, Cascini G, Fiorineschi L, Rotini F (2011) Dispositivo e apparecchiatura di triturazione e deumidifidicazione di materia prima vegetale. Patent Application # ITFI2011A000094
7. Andersson E, Harvey S, Berntsson T (2006) Energy efficient upgrading of bio-fuel integrated with a pulp mill. Energy 31(10–11):1384–1394
8. Wolf A, Vidlund A, Andersson E (2006) Energy-efficient pellet production in the forest industry—a study of obstacles and success factors. Biomass Bioenergy 30(1):38–45
9. Valentino GA, Leija L, Riera E, Rodriguez G, Gallego JA (2002) Wood drying by using high power ultrasound and infrared radiation. In: Forum Acousticum, Sevilla, 16–20 Sept 2002
10. Mujumdar AS (2007) An overview of innovation in industrial drying: current status and R&D needs. Transp Porous Media 66(1–2):3–18

Chapter 5
Application of IPPR to the Reengineering Problems of Class 2

5.1 Introduction: The Italian Accessible Fashion Footwear Industry

The methodology has been applied to a branch of the Italian footwear industry that has strongly contributed to the national industrial growth in the second half of the twentieth century. The sector played a significant role in the success of products marked "Made in Italy", synonym of prestige and glamour, but is actually facing a crisis period, although the style of the manufactured shoes is still considered original and fashionable. Particular difficulties are encountered by shoe factories and industrial districts producing stylish and high-quality items, but not possessing the assets for recruiting top fashion gurus.

More in detail, the Italian footwear sector comprises a group of famous brands and a multitude of so called "accessible fashion" factories. The brands belong mostly to luxury market segment and determine or somehow considerably impact the emerging fashion trends. The high fashion industries show a good market share, since their success is based on customers' identification with the brand. The products of the accessible fashion enterprises address the consumers of a large income bracket. These firms cannot base their competition on a strategy swiveling on low selling prices, due to the higher margins displayed by industries of the emerging countries, whereas the labor is much cheaper. Among the others, the worldwide market of discount shoes is dominated by Asian products and the economical trends highlight further reductions of Western productions in this business. On the other hand, the biggest majority of accessible fashion factories cannot aspire to enter the high-end market, because they miss the means, the organizational structure and the know-how of the brands.

In such context, IPPR was applied with the aim of analyzing the whole business process of the firms facing market difficulties, in order to remark the greatest value bottlenecks and the most promising opportunities for improving the competitiveness.

F. Rotini et al., *Re-engineering of Products and Processes*,
Springer Series in Advanced Manufacturing, DOI: 10.1007/978-1-4471-4017-7_5
© Springer-Verlag London 2012

The following Section describes more in detail the problems faced by the accessible fashion firms, underlining how the display of the business process contributes to the current crisis. Subsequently, Sect. 5.3 illustrates the application of IPPR leading to the comparative value analysis among the business process phases in charge of the shoe factories. Eventually, Sect. 5.4 presents a brief discussion on the obtained outcomes with the aim of highlighting their robustness and effectiveness within a reengineering initiative.

5.2 General Overview of the Business Process

The yearly activity of the footwear industry is mainly based on two market seasons (summer and winter), resulting in a strong influence on the organization of production and manufacturing activities. The performed business process may be subdivided in three main blocks of phases, briefly described in the followings.

The seasonal process begins with the realization of a big amount of prototypes. Such stage is aimed at providing a collection of samples according to style and relevant features of the shoes that are attributed by fashion designers on the basis of the vogue trends. The factories sell their products on the basis of these samples, in the shape of three-dimensional models, previously submitted to several tests. Among the outputs of the prototyping, a very important issue is the definition of the bill of materials for each shoe model included in the collection that is required for the scheduling of the manufacturing process, the purchases planning and the determination of the prices.

After the prices are set on the basis of the expected cost and potential commercial success of the shoes, the seasonal offer of the firm is fixed. This enables the starting of the selling stage, commonly carried out by agents through the intensive participation to sector fairs. In this phase the factories receive selling orders for the presented shoes, on the basis of which they plan the manufacturing activities. During this stage the agents and the sellers continuously update the factory about the sold batches, so that the scheduling is readjusted according to more and more reliable forecasts.

Subsequently, once the initial planning of the purchases and the manufacturing activities is completed, the firm keeps in coordinating the shoes production as the process progresses. The manufacturing manager and the purchases responsible are in charge of continuously supervising the accuracy and the timeliness of the operations. The shoes manufacturing is constituted by several sub-tasks, such as production of working tools (i.e., dies and shoe lasts), acquisition of leather, heels and components, uppers manufacturing and sewing, assembling etc. Due to economic convenience, most of the listed activities are carried by subcontractors (usually both offshore and onshore). The batches of shoes are then shipped to the retailers, representing the direct customers of the factories. The enterprise has to put attention on the flawed products along the whole manufacturing stage. Figure 5.1 provides a graphical representation of the main activities involved in the classical business process in charge of the shoe factories.

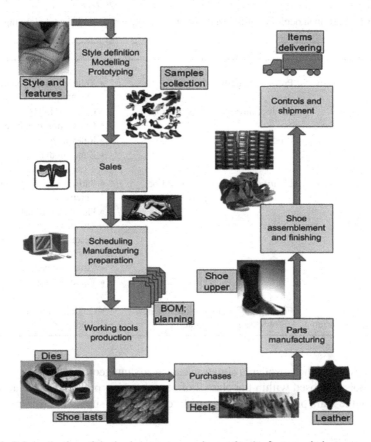

Fig. 5.1 Schematization of the business process relevant for the footwear industry

Due to the characteristics of the business process, the collection of the shoe samples must-be prepared substantially in advance with respect to the buying of the end-users. In the meantime, the powerful influence of the brands induces strong changes in the vogue trends, potentially resulting in a mismatch between the tastes of the final consumers and the stylistic features of the items produced by many shoe factories. This generates large amounts of unsold goods coming from the accessible fashion industry, leading to retailers' dissatisfaction towards these firms and influencing negatively the repurchase intentions. As a consequence, the accessible fashion firms lose relevant market shares.

Hence, on the basis of the previous considerations, the shoe factories belonging to the accessible market suffer of a lack of competitiveness, since the provided output, although already sold, is not able to fulfill the style expectations of the end-users.

As a result, the present business process requires a reengineering strategy to fill the gap between the offered product features and the wishes of the final consumers that are influenced by the emblazoned labels. Given the mismatch between the industrial process and the expected outputs, resulting in a general lack of

Table 5.1 IPPR methodology customized for the reengineering problems belonging to the class 2

Phase	IPPR activity	Tools
Step 1		
Process to problem	Process modelling	• Multi-domain modeling technique
	Product information elicitation	• CRs checklist
		• Correlation coefficients
	Product modeling	• Relevance scale
		• Kano model
Step 2		
Problem to ideal solution	Identification of what should be changed in the process	• Satisfaction/dissatisfaction metrics
		• Phase overall satisfaction metric
		• Resources consumption metric
		• Value indexes
		• Value assessment chart
Step 3		
Ideal solution to physical solution	Finding physical solutions for new process implementation	• Guidelines to select process redesign tools

competitiveness, the considered situation can be advantageously analyzed by means of the tools proposed within the class 2 of reengineering problems. This allows to point out the main process criticalities with regards to the supply of customer value.

5.3 Application of IPPR

The aim of the present Section is to describe the application of IPPR in the version tailored to treat the reengineering problems falling into the class 2.

The step by step sequence of activities, recalled in Table 5.1, leads from the segmentation of the process to the individuation of the value bottlenecks and the consequent favorable directions for the business reorientation.

5.3.1 Process to Problem

The first activity of IPPR requires the modeling of both the industrial process and its outputs in terms of relevant product features. The multi-domain modeling technique has been employed to build the functional scheme of the business process. Hence, the product representation has been carried out by means of the Kano model and the relevance scale, revealing the role played by each customer requirement in determining the satisfaction for the buyer. Finally, the relationships between the phases and the delivered product features have been investigated and expressed through the correlation coefficients, in order to subsequently estimate the contribution of each process segment in the determination of the whole value.

5.3.1.1 Process Modeling

As previously recalled, the companies belonging to the accessible fashion footwear industry are characterized by production processes sharing considerable commonalities. Therefore, the specific scope of IPPR employment has resulted in the attainment of strategic directions for the evolution of the whole sector rather than for a single firm. According to this aim, the building of a reliable multi-domain model of the business process has required a deep exploration of the sector, with a particular effort dedicated to individuate values of the involved coefficients representative for the whole set of investigated firms. More specifically, the presented analysis refers to a large sample of firms composing an industrial footwear district in central Italy.

At first, the information referable to the sketch of the business process presented in Sect. 5.2, has been enriched through the consultation of technical publications in the field. Such an activity has allowed the subdivision of the process into the set of relevant phases, then confirmed throughout the subsequent information gathering tasks. The segmentation of the process and the individuation of the main flows has been performed by using the criteria and the formalisms of IDEF0 model (recalled in the Appendix A), as suggested in the IPPR roadmap. Figure 5.2 illustrates the sectioning of the industrial process into three main blocks of activities and depicts, although not in charge of the shoe factory, also the retailing phase, which produces relevant feedbacks on the future collections. The three principal blocks are further partitioned, giving rise to the schemes of Figs. 5.3, 5.4, 5.5.

According to the above representation, the business process can be segmented into the following main phases, whose label is in brackets:

- the process starts with the determination of the proper footwear style in charge of experts which analyze the trends of the big firms and fine-tune the specifications for the seasonal collection (A11);
- the collection is designed (A12);
- the samples of the shoes are produced, as well as the technical documentation is provided in order to ease the subsequent schedule of the manufacturing (A13);
- the selling price is established for each kind of shoe and the shoe factory participates to the main sector fairs, whereas the negotiations with interested retailers take place (A21);
- the selling agents carry on collecting the orders (A22);
- according to the received orders, the manufacturing operations and the purchasing of the needed components are planned (A31);

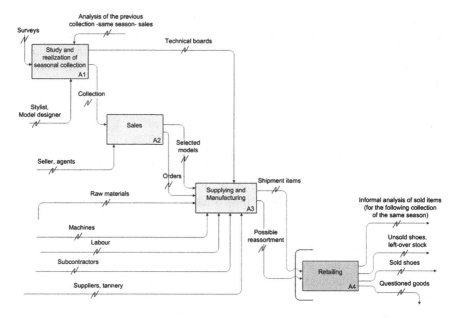

Fig. 5.2 Layout of the business process relevant for the shoe factories, operating in the accessible fashion bracket, represented through IDEF0 model

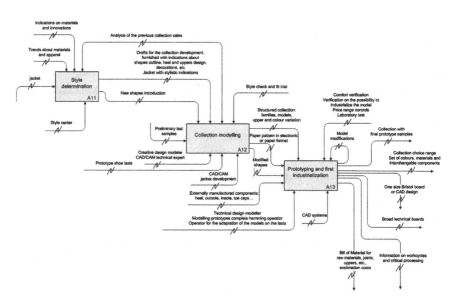

Fig. 5.3 Segmentation of the block of activities, aiming at the ideation, design and prototyping of the shoes collection through IDEF0 model

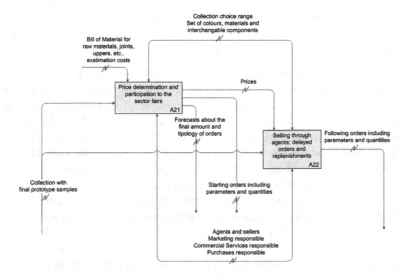

Fig. 5.4 IDEF0 scheme of the main phases addressed at the selling the designed shoes collection

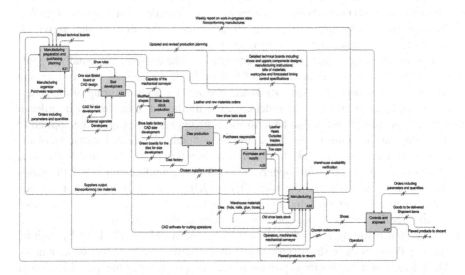

Fig. 5.5 IDEF0 model of the operations addressed at finalizing the shoes production according to the previous phases

- the different sizes are developed with regards to each sold item belonging to the collection (A32);
- the shoe lasts are manufactured by specialized parties or units (A33);
- the dies for the cutting of the different parts of the shoe are produced (A34);
- the purchasing of the needed parts is carried out (A35);

- the seasonal batch of shoes is manufactured (A36);
- the produced items are verified in order to check their compliance with the standards of quality and the shoes are shipped to the retailers (A37).

The second stage of the analysis has been supported by a group of experts, which was constituted by an analyst having 20 years of practice in the field and three entrepreneurs. The collaboration of the sample of specialists was aimed at gaining further understanding about the manufacturing activities, the involved organizational skills and the relevant industrial practices. As a result of the additional information, some key phases have been further segmented. However, according to the scope of representing a business condition relevant for a whole sector, it emerged that the individuated phases could not be further characterized by indexes on which a considerable sample of firms could converge. In other words, the characterizing coefficients of the individuated segments, within IPPR, present an extreme variability according to each single enterprise. Indeed, for instance, the employment of the resources in the sub-operations is strongly dependent on the specific company. The kind of shoes (classical shoes, sandals, boots, moccasins etc.), the characterizing stylistic features (use of accessories, presence of decorations and seams), the reference markets (South Europe, North Europe, USA, Russia, Japan etc.), the gender of the end-users, the collection (summer or winter) heavily influence the process practices, resulting in noticeably different use of resources (financial commitments, duration of the activities, employed labor, carefulness in operations performing.

For the sake of completeness, the acquisition of the components usually regards the purchase of the required amount of leather, heels, soles, shoe tips, insoles and additional accessories, e.g. zips, studs and clasps. The manufacturing of the shoes, according to the design and the orders, starts with the cut of the leather in order to obtain the required pieces that have to form the uppers, the strips to be applied to the heels and the hemlines for the soles. The suitable shapes and measures of the cut leather are obtained thanks to the prior preparation of the sizes and of the dies. The parts composing the shoe uppers are sewn by the binding unit, while the bands are joined to the heels and the soles. The purchased parts and the semi-finished components are assembled and shaped throughout the shoe lasts. At last the shoes are finished by means of the operators working at the conveyor.

The final resulting multi-domain model has been definitively built by visiting the shop floors of three shoe factories and by consulting their production managers. The refinement of the business process model has given rise to the quantification of the channeled resources, assessing the ordinary elapsed times and ratios of monetary expenditures addressed to each phase.

The last investigation stage has indeed allowed to obtain further relevant indications, such as:

- common problems faced in the business planning;
- constraints related to the manufacturing process (e.g. common production capacity of the devices, sequence of the operations to be performed, rules to be followed);

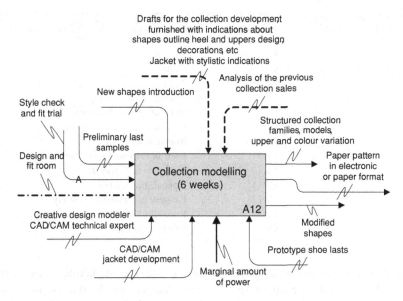

Fig. 5.6 Schematization of the phase A12 of the business process through the multi-domain model proposed within IPPR

- extent of the involved resources in terms of incurred costs;
- employed personnel;
- common duration of the phases.

According to the gathered information, the business process has been schematized by means of the proposed multi-domain model. An exemplary phase is reported, as illustrated in Fig. 5.6, while the whole schematization is omitted for space reasons.

5.3.1.2 Product Information Elicitation

The commonalities existing in the investigated sector allow to provide a shared representative scheme of the process output. The activity illustrated in the present paragraph brings to the definition of the main product features and the estimation of the phases contribution to their fulfillment, throughout the consultation of the recalled sector experts involved in the reengineering task.

The determination of the relevant customer requirements has taken into account the fact that the process operates in a B2B environment. Therefore, the relevant product attributes deal with both the direct customer (i.e. the retailer, the outlet) and the end-user of the shoes, which follows in the value chain. The record of customer requirements has been extracted by resorting to both the process model and the CRs checklist. In the followings, some examples are provided to illustrate the elicitation process of the product features playing any role in the delivery of satisfaction.

CR	Business process attributes
CR1	Customization possibility
CR2	Link with the apparel sector
CR3	Comfort
CR4	Technical and healthy properties
CR5	Standard sizes supply
CR6	Resistance and duration
CR7	Components completeness
CR8	Manufacturing care
CR9	Care in the order dispatching
CR10	Non-standard sizes availability
CR11	Appeal, lines, shapes
CR12	Colours and materials variety
CR13	Possibility of developing faithful customers
CR14	Offer potentiality

Table 5.2 Set of relevant customer requirements for the business process of the footwear industry

With reference to the customer dimension and the marketplace, the sector experts have individuated relevant competing factors, which concern, more specifically, the *carefulness in dispatching the orders* (then labeled as CR9) and the *variety* (CR12) of the shoes belonging to the seasonal collection. On the other hand, several customer requirements regard the sphere of the end-user and emerge by analyzing the business process model and focusing on the transformations occurring along the phases. For instance, the way the collection modeling is performed, including a preliminary testing of the prototype footwear, leads to the individuation of the shoes *comfort* (CR3) as a significant competing factor. Additionally the consultation of the CRs checklist has allowed the identification of further attributes, which were previously neglected, e.g. the term "the opportunity provided to advantageously employ the product for not standard users or disabled people" has given the chance to consider the value exerted by ensuring the capability to provide *non-standard sizes* (CR10) of the shoes.

The mechanism for the elicitation of the product attributes has given rise to the list of customer requirements depicted in Table 5.2, subsequently employed for the purpose of the IPPR application. It is worth to notice that the number of customer requirements (14) is similar to that of the investigated phases (12); thus, the analysis respects the indication to employ a ratio between these two quantities ranging from 1/2 to 2.

Most of the customer requirements are not referable to quantitative and measurable parameters, nor they can be considered as a result of one or more engineering characteristics. The determination of the correlation coefficients has then involved insightful considerations of the sector experts, in order to converge towards a shared scheme of the phases contributions in attaining the customer requirements. Two examples are reported with the aim of elucidating the reasoning carried out by the specialists to indicate the extent of the k_{ij} indexes. For instance, the *link with the apparel sector* (CR2) is determined just by the expected capability of the model designer to catch the tendencies taking place in the clothing industry;

Table 5.3 Correlation coefficients pertaining the link among the phases of the business process and the customer requirements to be fulfilled

CR / Phase	1	2	3	4	5	6	7	8	9	10	11	12	13	14
A11	0	1	0	0	0	0	0	0	0	0	0.7	0	0.7	0
A12	0	0	0.2	0.2	0	0.2	0	0.2	0	0	0.05	0	0	0
A13	0.1	0	0.2	0	0	0	0	0	0	0	0.25	0.4	0	0
A21	0	0	0	0	0	0	0	0	0.05	0	0	0	0.15	0.5
A22	0	0	0	0	0	0	0	0	0.1	0	0	0	0.15	0.5
A31	0.5	0	0	0	0	0	0.6	0	0.5	0	0	0.4	0	0
A32	0	0	0	0	0.6	0	0	0	0	0.6	0	0	0	0
A33	0	0	0	0	0.2	0.1	0	0.1	0	0.2	0	0	0	0
A34	0	0	0	0	0.2	0	0	0.1	0	0.2	0	0	0	0
A35	0.1	0	0.3	0.4	0	0.3	0.2	0.1	0	0	0	0.1	0	0
A36	0.3	0	0.3	0.4	0	0.4	0.2	0.5	0.05	0	0	0.1	0	0
A37	0	0	0	0	0	0	0	0	0.3	0	0	0	0	0

as a consequence the *style determination* phase (A1) is completely in charge of the fulfillment of such requirement and $k_{CR2-A11}$ assumes the value 1. Conversely, the CR9, i.e. *care in the order dispatching*, is influenced by manifold factors intervening along the business process. At first, in order to pursue the correctness of the order, the personnel employed in the selling stages has to carefully record and communicate the details related to the job (quantities, typologies, colors, further finishes, etc.); phases A21 and A22 (at a greater extent due to the bigger number of negotiations) are then relevant for the CR9. A major contribution to the display of such customer requirement is provided by the phase A31, aimed at organizing the whole production process, planning the sequence of purchases and the manufacturing activities, checking and updating the job status for each customer. The manufacturing (A36) is responsible for the care dedicated to the orders, in terms of respecting the assigned specifications. At last, the finalization of the orders takes place in the last phase (A37), whereas any mistake can be revealed, giving rise to further rescheduling of the wrong jobs and re-working of the flawed items. The correlation coefficients related to CR9, as resulting from the above discourse, are showed in the pertaining column of Table 5.3, which summarizes all the fractions addressed to each phase in ensuring the achievement of the customer requirements, i.e. the k_{ij} indexes.

5.3.1.3 Product Modeling

The identification of the process bottlenecks affecting the delivery of value is performed on the basis of the phases' contribution in generating the satisfaction and in avoiding the dissatisfaction of the customer. Therefore, beyond the assessment of the relevancies in impacting the perceived value, also the classification of the customer requirements according to the Kano model has been carried out, as foreseen for the second class of reengineering problems. Due to the lack of recent

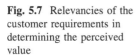

Fig. 5.7 Relevancies of the customer requirements in determining the perceived value

and reliable customer surveys, the product modeling task has been carried out by employing the opinions and the evaluation of the sector experts.

The specialists have suggested, on the basis of a 1–5 Likert scale, the degrees of relevance summarized in the diagram of Fig. 5.7.

As shown, the possibility to operate the customization of the product (CR1) in addition to the completeness of the shoe in terms of components and accessories (CR7), the attention in fulfilling the order (CR9), the aesthetic features characterizing the style (CR11) and the availability of different colors and materials (CR12), are the characteristics having the major impact on the perceived value.

The classification of the customer requirements according to the Kano model has been performed by fostering the experts reflections by means of the algorithm suggested in Chap. 3.2.3.2. Hereinafter, an example is provided with the aim of clarifying its application.

Let's consider the CR1—Customization possibility, the following question has been submitted to the experts:

Does the improper fulfillment of the Customization possibility provoke customer dissatisfaction?

They answered "No", since it has been deemed that such characteristic is not expected by the end-user when planning to buy new shoes. Therefore, according to the flow of questions foreseen in the sequence, the experts were further consulted about the following issue:

Does the correct accomplishment of the Customization possibility contribute to customer perceived satisfaction?

The possibility to customize certain details of the shoes has been considered as an unspoken feature impacting positively the satisfaction of the customer; thus they answered "Yes" to the question. According to the response, the CR1 has been classified as an *Attractive* characteristic.

The classification algorithm, applied together with the sector experts, brought to the categorization of the customer requirements, as shown in Table 5.4.

5.3.2 Problem to Ideal Solution

In this section, the determination of the value indexes characterizing each process phase is presented. Thanks to the correlation coefficients, the calculation of the contribution of each phase in generating the customer satisfaction and in avoiding

Table 5.4 Classification of the customer requirements according the Kano model

Customer requirement	Kano category
CR1	Attractive
CR2	Attractive
CR3	Must-be
CR4	Must-be
CR5	Must-be
CR6	Must-be
CR7	Must-be
CR8	Must-be
CR9	Must-be
CR10	One-dimensional
CR11	One-dimensional
CR12	One-dimensional
CR13	Attractive
CR14	One-dimensional

the customer dissatisfaction has been carried out. The resources consumption index expressing the amount of resources involved by each phase has been evaluated. Subsequently, the Overall Value, the Value for Exciting requirements and the Value for Needed requirements have been assessed giving rise to the identification of the required reengineering actions.

5.3.2.1 Phase Overall Satisfaction

The assessment of the contribution of each process segment to the customer satisfaction arises as the mathematical formulas introduced in Chap. 3.3.1.3 are applied. The previously collected indexes allow the determination of the *Customer Satisfaction* (CS) and *Customer Dissatisfaction* (CD), according to the followed expressions, here reported for the sake of simplicity:

$$CS_i = \frac{o_i + a_i}{A + O + M}; \ CD_i = -\frac{m_i + o_i}{A + O + M}.$$

The resulting values are summarized in the Table 5.5.

Subsequently, thanks to the correlations coefficients expressing the relationships between each phase and customer requirement, the contribution of each process segment in determining the customer satisfaction and in avoiding the dissatisfaction, has been calculated by means of the following relations:

$$PCS_j = \sum_i k_{ij} \times CS_i; \ PCD_j = \sum_i k_{ij} \times CD_i$$

Table 5.5 Customer satisfaction and dissatisfaction coefficients assessed for each customer requirement

CR	Kano category	a_i	o_i	m_i	CS_i	CD_i
CR1	Attractive	4	0	0	0.10	0.00
CR2	Attractive	3	0	0	0.08	0.00
CR3	Must-be	0	0	2	0.00	−0.05
CR4	Must-be	0	0	1	0.00	−0.03
CR5	Must-be	0	0	2	0.00	−0.05
CR6	Must-be	0	0	1	0.00	−0.03
CR7	Must-be	0	0	5	0.00	−0.13
CR8	Must-be	0	0	2	0.00	−0.05
CR9	Must-be	0	0	4	0.00	−0.10
CR10	One-dimensional	0	1	0	0.03	−0.03
CR11	One-dimensional	0	5	0	0.13	−0.13
CR12	One-dimensional	0	4	0	0.10	−0.10
CR13	Attractive	3	0	0	0.08	0.00
CR14	One-dimensional	0	3	0	0.08	−0.08

Table 5.6 Phase contributions in determining the overall satisfaction

Phase	PCS_j	PCD_j	POS_j	$POS_j(\%)$
A11	0.22	−0.09	0.12	16.90
A12	0.01	−0.04	0.03	4.23
A13	0.08	−0.08	0.08	11.27
A21	0.05	−0.04	0.04	5.63
A22	0.05	−0.05	0.05	7.04
A31	0.09	−0.17	0.15	21.13
A32	0.02	−0.05	0.04	5.63
A33	0.01	−0.02	0.02	2.82
A34	0.01	−0.02	0.02	2.82
A35	0.02	−0.07	0.06	8.45
A36	0.04	−0.10	0.08	11.27
A37	0.00	−0.03	0.02	2.82

Hence, the overall satisfaction generated by each process segment has been assessed through:

$$POS_j = 0.29 \times PCS_j - 0.04 \times PCS_j^2 - 0.72 \times PCD_j + 0.07 \times PCD_j^2$$

The obtained results are shown in Table 5.6. Two main phases stand out as crucial in giving rise to the customer contentment, i.e. the Style determination (A11) and the Manufacturing preparation and purchasing planning (A31). Conversely, all the other phases have a consistently lower impact on the overall satisfaction.

5.3.2.2 Resources Consumption

The criticalities of the considered business process essentially depend on issues related to the required time to supply the finished product and to the expenditures occurring to acquire the needed resources. The experts have stated a similar influence of time and cost factors with regards to the role played to the detriment of the sector competitiveness. Along the display of the industrial process, the emergence of harmful functions is conversely limited (e.g. the machinery presents rare failures) and the sensitivity is poor for the competing problems with respect to such undesired effects. Such an evidence suggests to neglect the arising unwanted phenomena in the computation of the Resources consumption index, calculated with respect to a summer seasonal collection of shoes.

The latter, according to the above considerations, has been calculated by summing up the shares of costs and elapsed times related to each process segment, by means of the following formula:

$$RES_j = C_j + T_j.$$

The terms C_j and T_j represent the portion of costs and times accounted to the phases with respect to the whole expenditure and duration of the business process. They are computed as follows, giving rise to the values summarized in Table 5.7:

$$Cost\,rate\,of\,j\text{-}th\,Phase = \frac{Costs\,of\,the\,j\text{-}th\,Phase}{\sum_{j=1}^{N} Phase\,costs}$$

$$Time\,rate\,of\,the\,j\text{-}th\,Phase = \frac{Elapsed\,time\,of\,the\,j\text{-}th\,Phase}{\sum_{j=1}^{N} Elapsed\,times\,of\,the\,Phase}$$

More specifically, in order to determine the illustrated extents, the costs have included the consideration of acquired materials, labor, auxiliary operations, energy, amortization of the machinery, management of the firm. The computation of the expenditures has disregarded the costs ascribable to real estates, given the poor contribution in the overall amount of costs. The duration of the phases has considered the period of time elapsed between the beginning and the conclusion of the involved activities. Such choice has thus allowed to take into account dead times and the actual influence of the phases in delaying the supply of the products to the retailers, hence the loss of the fashion content of the shoes, as well as the consequent risks.

As shown by the outcomes, the Manufacturing (A36) and Purchases and supply (A35) phases involve a high consumption of resources. Although at a smaller

Table 5.7 Resources consumption index of each process phase

Phase	Name	Cost rate	Time rate	RES_j	$RES_j(\%)$
A11	Style determination	0.01	0.06	0.07	3.48
A12	Collection modelling	0.02	0.05	0.07	3.48
A13	Prototyping and first industrialization	0.02	0.05	0.07	3.48
A21	Price determination and participation to the sector fairs	0.01	0.03	0.04	1.99
A22	Selling through agents; delayed orders and replenishments	0.09	0.06	0.15	7.46
A31	Manufacturing preparation and purchasing planning	0.03	0.17	0.20	9.95
A32	Size development	0.00	0.09	0.09	4.48
A33	Shoe lasts stock production	0.01	0.03	0.04	1.99
A34	Dies production	0.02	0.09	0.11	5.47
A35	Purchases and supply	0.36	0.15	0.51	25.37
A36	Manufacturing	0.42	0.15	0.57	28.36
A37	Controls and shipment	0.01	0.08	0.09	4.48

degree, also A22, A31 and A34 present a not negligible commitment of times and costs, whilst the remaining phases result definitively less expensive.

5.3.2.3 Overall Value and VAC Diagram

Through the relationships presented in the previous Chapter (3.3.1.4–3.3.1.5), the value indexes that express the phases' performance compared with the employed resources have been calculated.

The global efficiency of the phases is evaluated by the means of the *Overall Value* (OV) and additional information about the process segments are obtained through the *Value for Exciting requirements* (VE) and *Value for Needed requirements* (VN). The last two mentioned indexes point out the phases' contribution to achieve delighting and basic product properties, respectively. All the recalled parameters employed for the value assessment are summarized in Table 5.8.

Finally, Fig. 5.8 shows the *POS versus RES Assessment Chart* (PRAC), which compares the scores of the provided satisfaction and of the global resource channeling. The *Value Assessment Chart* (VAC), reported in Fig. 5.9, has been the basis for the following discussion about the reengineering priorities. Being the number of the phases quite restricted, the means of VE and VN coefficients (both around 0.6) have been chosen to discriminate between low and high values and, therefore, to subdivide the diagram into the four performance areas. In Fig. 5.10 the low performance area is zoomed.

The Overall Value index shows that the main critical issues concern the phases dealing with the manufacturing, the supply chain and the working tools production. These phases, that take place after the engineering of the collection, represent bottlenecks in the value chain creation process, showing a growth in the employed resources.

Table 5.8 Value indexes characterizing the shoes manufacturing process

Phase	Name	OV	VE	VN
A11	Style determination	1.88	3.26	1.33
A12	Collection modelling	1.25	1.23	1.23
A13	Prototyping and first industrialization	1.24	1.35	1.17
A21	Price determination and participation to the sector fairs	0.73	0.45	0.82
A22	Selling through agents; delayed orders and replenishments	0.43	0.12	0.54
A31	Manufacturing preparation and purchasing planning	0.43	0.09	0.55
A32	Size development	0.41	0.16	0.49
A33	Shoe lasts stock production	0.32	0.32	0.31
A34	Dies production	0.24	0.00	0.33
A35	Purchases and supply	0.15	0.07	0.17
A36	Manufacturing	0.14	0.04	0.18
A37	Controls and shipment	0.12	0.04	0.14

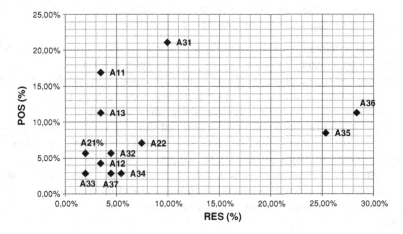

Fig. 5.8 The PRAC diagram which compares the provided overall satisfaction with the global resource channeling

The charts of Figs. 5.8 and 5.9 show that several phases belong to the low performance area. Conversely, few phases are tailored to ensure both the basic properties and those product attributes which are unexpected and generate a higher level of customer value.

The phase aims at defining the style of the shoes line is marked by High Value, but the fulfillment of exciting features is strongly predominant. Only manufacturing preparation and purchasing planning phases belong to the Basic Value area; this stage is mainly constituted by management activities and control operations aimed at dispatching correctly the orders. Among the phases in Low Value area, the ones with the worst value indexes are characterized by prevalent orientation towards the necessary features and by large resources utilization; the dies production process shows even limited benefits.

Fig. 5.9 Value assessment chart of the process built through VE and VN coefficients. The four performance areas have been identified through the average values of the indexes (around 0.6)

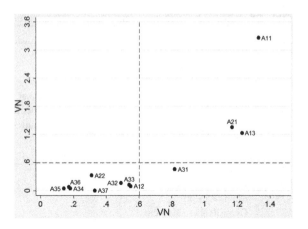

As a consequence, on the basis of the meaning attributed to the four value areas, the following evidences about the business process have emerged:

(1) the phases belonging to High and Basic Value areas (A11, A13, A21 and A31) have to be safeguarded, therefore they do not require substantial changes;

(2) the position of the phases A12, A32 and A33 is doubtful, straddling the low and basic benefits areas; in each case, notwithstanding their low value, these operations are not the ones requiring the greatest attention;

(3) the phase A34 regarding dies production and showing very low benefits and a not negligible consumption of resources, is worth to be trimmed or however to be submitted to major technological changes;

(4) the phase A37 concerning controls and shipment, shows the lowest share in the attainment of customer satisfaction and it is marked by the worst OV index, although contributing not marginally to the determination of expected product features;

(5) the phase A22, selling through agents, is situated in the centre of the Low Value area and it is the only one showing the opportunities to jump in the region of Exciting Value;

(6) the phases A35 and A36, concerning the purchases and the manufacturing activities, belong to the Low Value area; besides showing a meaningful contribution to pursue the basic product features, the reason of the unsatisfying performances resides in the high score of resources consumption (as remarked in Fig. 5.8), thus the reengineering actions should be oriented towards time and money savings;

(7) no phase is currently situated in the Attractive Value region and there's a need to investigate the emerging trends and the successful issues in the footwear sector, thus leading to revise the business process phases or even to add new ones.

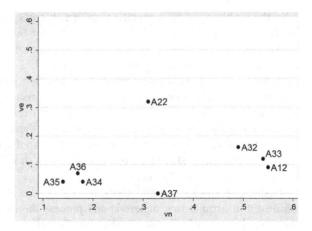

Fig. 5.10 The low performance area of the Value Assessment Chart

5.3.3 Ideal Solution to Physical Solution

The main implications of the analysis consist in the adoption of reengineering measures pertaining both the whole process and single phases. On the one hand, the poor performance of complex and basilar process segments, involving the supply and the manufacturing of the materials, determines the need to advance consistent redesign efforts, especially with the aim of reducing the resources consumption. Besides, the lack of efficient phases specifically tailored to deliver delighting product features suggests the individuation of new attributes. On the other hand, several bottlenecks affecting the process emerge as a result of the study of the business process according to value terms. The analysis of the criticalities performed through the OV index and the VAC diagram advocates prioritizing the reengineering efforts on the phases A34 and A37, beyond the already mentioned A35 and A36, which are worth to be considered within a wider perspective. According to further insights and due to limited benefits, the examination has revealed the need to verify the opportunity to migrate the functions carried out by "dies production" and "controls and shipments" phases.

5.3.3.1 Directions for a Global Process Rethinking

As previously recalled, the extent of the phases A35 and A36, showing unsatisfying value indexes, suggests to face the reengineering problem in terms of an overall task of process redesign. The results of the investigation shed light to criticalities related to the cited process segments with reference to the treated sector; however, the complexity of the involved operations and the marked differences among the firms suggest to carry out further value analyses of the activities according to the specific purchasing and manufacturing operations. Given the constraints of the manufacturing stages, intended to fulfill the scheduled jobs, the problems of each specific firm or unit can be advantageously faced with the IPPR instruments advised for the first class of business reengineering tasks.

Nevertheless, measures to reduce the amount of channeled resources could result strongly beneficial, regardless the peculiarities of each enterprise. Upstream activities, such as the organization of the production, operate in order to minimize the costs of the raw materials and of the semi-finished items. In each case, such issue does not result sufficient to align the spent resources to the generation of customer satisfaction. In this sense, according to the schema proposed in Chap. 3 for the identification of suitable reengineering tools, the employment of Lean Manufacturing hints should be evaluated by the shoe factories belonging to the sectors and by the supply chain.

Besides, if any meaningful reduction cannot be pursued of the expenditures addressed to acquire raw materials and process the required components, due to the limited sphere of influence of the shoe factories, a reduction of time resources should be also considered. More specifically, the manufacturing phase is fragmented and the coordination of its inherent activities results difficult and time consuming. Major troubles take place especially when the participation of offshore outsourcers occurs, because of transportations and exponentially increasing delays in case of mistakes or flawed semi-finished goods to be reworked or remade.

The possibility to reduce time resources would allow strategic phases (primarily the discussed activities related to purchases and manufacturing) to be repositioned. In order to reach this goal, dead times have to be strongly shortened. According to the suggestions provided in the third IPPR step for process reengineering, the means of *Quick Response* strategies result the most appropriate. The implementation of the recommended measures would imply the minimization of lead times and the consequent reduction of time to market. From this perspective, the analysis of existing technologies, policies and methods provides precise indications about the actions to be taken. According to the quick response criteria, the firms can carry out the purchases of the materials with a higher supplying time (especially leather), before the sales stage and on the basis of a forecasted quantity with a safety margin; this is however possible just with the products that can be reused, recycled or reworked, unless the firm accepts high entrepreneurial risks. Besides, the engagement of offshore subcontractors should be limited in order to reduce the phase duration and to ensure the timeliness of the commitment; moreover, researches show that onshore manufacturing and the creation of domestic partnerships are more profitable for a certain share of produced items [1].

The process shortcomings related to the miss of a sufficient amount of delighting features regards the sphere of the product and could be advantageously faced by the means of IPPR tools introduced for the third class of reengineering problems. Some evidences emerging from literature sources are briefly discussed in Sect. 5.4.

5.3.3.2 Measures to Overcome Process Bottlenecks

The arising indications which regard the phases A34 and A37 suggest their suppression and the implementation of their performed functions to other portions of the process.

The search of more suitable ways for leather cutting, replacing the production of the dies, suggests the employment of CAD/CAM systems and automated machines. Such modifications would involve consistent transformations in the sizes development phase, needing to implement new technologies, allowing in addition to augment the delivered value. Similar solutions are already adopted by similar firms of the same industrial sector, not belonging to the investigated district and thus not under examination.

The process phase in charge of controls and shipments is difficult to be substituted, unless severe transformations in the organization of the business process are applied. In other industrial contexts warehousing technologies hold the capability to assist the managing of the performed functions, allowing the A37 phase to become more agile, if not completely trimmed and integrated in the manufacturing finalization.

5.4 Discussion of the Outcomes

Among the possible approaches to reverse the negative trends in the accessible fashion footwear industry, the quick response strategies are aimed at following the rapid changing market scenarios. They individuate the speed as primary competing factor for boosting the success of the enterprises, as confirmed by plenty of literature contributions and experiences. Such practices arose during the 1980s in the apparel sector [2, 3], that has first expressed the need to speed up the introduction of the produced items in the marketplace. Minor efforts have been dedicated to the adoption of quick response policies in the footwear industry, although resulting in remarkable benefits [4], especially whereas the whole supply and retailing chain has been involved.

Recently, the urge to undertake quick response strategies as a potential help for fashion shoe factories has been the focus of a project named Just In Time for Shoes (JITS). The latter, funded by Tuscany Region and coordinated by PQuadro (a consultancy society with a long lasting experience in the footwear industry) has analyzed the competitive situation of several factories belonging to the district, which was also the object of the present investigation. As emerging by the outcomes of the project, a successful application (although partial) of quick response strategy was effectively carried out by one of the examined enterprises. Such practices have been introduced by a small shoe factory that operates just in the foreign market and is specialized in the middle quality women shoes segment, although other kinds of items are produced. The production of the firm is characterized by limited lead times considering the average elapsed duration between the orders acquisition and dispatching (about 60 days, while commonly the current business process employs between 3 and 4 months), the elimination of dead times, continual replenishments, offshore outsourcing applied just for few low quality items, anticipated purchase of the leather, improved coordination of the supply chain. Thanks to these business choices, the shoe factory has not suffered the crisis in the sector and, unlike the general trend, the turnover of firm has almost doubled after the attainment of such measures.

Eventually, with reference to the attempt of identifying new attracting requirements, the main tendency in footwear industry, as documented in the literature, is related to the mass customization phenomenon. Already in the past years, a research [5] reveals that shoe consumers "are curious about the customization concept and do realize the related benefits"; moreover the potential success of customized fashion items is deliberately assessed [6, 7]. The capability of the current business process to operate the customization of few details of the shoes results insufficient, often representing a lure for the retailer, but with a limited impact on the end-user, ultimately contributing at a low extent to repurchase intentions.

References

1. Warburton RDH, Stratton R (2002) Questioning the relentless shift to offshore manufacturing. Supply Chain Manag Int J 7(2):101–108
2. Sullivan P, Kang J (1999) Quick response adoption in the apparel manufacturing industry: competitive advantage of innovation. J Small Bus Manag 37(1):1–13
3. Ko E, Kingade D, Brown JR (2000) Impact of business type upon the adoption of quick response technologies—the apparel industry experience. Int J Oper Prod Manag 20(9):1093–1111
4. Perry M, Sohal AS (2001) Effective quick response practices in a supply chain partnership—an Australian case study. Int J Oper Prod Manag 21(5–6):840–854
5. Piller FT, Müller M (2004) A new marketing approach to mass customisation. Int J Comput Integr Manuf 17(7):583–593
6. Neef A, Burmeister K, Krempl S (2005) Vom personal Computer zum personal fabricator (From personal computer towards personal manufacturer). Murmann Verlag, Hamburg
7. Franke N, Schreier M (2008) Product uniqueness as a driver of customer utility in mass customization. Mark Lett 19(2):93–107

Chapter 6
Application of IPPR to the Reengineering Problems of Class 3

6.1 Introduction: Overview of the Hairstyling Sector

The application of the instruments foreseen for the third class of reengineering problems potentially leads to disruptive innovations, regardless the treated product or business. In order to preliminarily test the adopted approach, a large amount of proposals have been advanced to the authors by SMEs, academics and students showing interest for an assisted or guided generation of creative business models and ideas, leading to numerous applications. Among the bundle of case studies, the generalities of the hairdressing sector have kindled a particular interest. The choice of attributing a particular attention to such field within the endeavor of experiencing the proposed IPPR tools has leant upon the highlighting of peculiarities about the minor modifications occurring in the hairstyling world, especially from the viewpoint of the ways the business is conducted.

The industry is strongly characterized by the presence of a multitude of small-scale hairdressing outlets and a consistently lower diffusion of salon chains and franchises of big companies [1]. The largest groups are progressively, although slowly, creeping in, also thanks to the organization of internal training schools; furthermore, being deemed to represent the glamorous branch of the industry, they are accounted to determine the tendencies about style and hairdressing techniques [2]. However, the main changes that have occurred along the last decades relate to external factors and mainly to demands of customers, especially women, looking for colors and styles requiring little time for home hair care. The increasing awareness of customers about style features and the consequent emergence of more sophisticated requests have brought to the need to enhance hairdressers professionalism and communication skills [3].

The described tendencies in the industry have poorly impacted the innovation of the equipment employed in the salons. The adaptation to trends in hair fashion can be satisfied with traditional tools, resulting in a slight rate of technological innovation [2]. The overwhelming diffusion of ICT, which has resulted as a common

F. Rotini et al., *Re-engineering of Products and Processes*,
Springer Series in Advanced Manufacturing, DOI: 10.1007/978-1-4471-4017-7_6,
© Springer-Verlag London 2012

feature for manufacturing industries and service retailers, has just marginally impacted the beauty sector, with an expected increased involvement that concerns just the booking and the management of the liaison with the customers [4]. The running innovations have therefore resulted in a negligible impact on the way hairdressers perform their main work (cutting, coloring, styling), although in face of requested outstanding capabilities. In this framework the hand-held hair dryer introduced in the past century can still be considered the main technological breakthrough [3]. The scientific literature providing insights about the working conditions in hairdressers shops is mostly dedicated to discuss the concerns about stylists health and safety of the salon environment. However, the argument has slightly impacted the innovation patterns pertaining the sector.

With these premises, the employment of reengineering tools to achieve new business opportunities in such a conservative industry has resulted a challenging task, suitable to verify whether the emerging ideas for innovation could stimulate the change also in the life of traditional hairdressing outlets. To this aim Sect. 6.2 provides additional information about the blow dryers employed in the hairdressing industry, which will be treated as an application of product reengineering with IPPR, as illustrated in Sect. 6.3. Eventually, the Chapter closes with Sect. 6.4, dedicated to discuss the outcomes emerged as a result of the application of the methodology.

6.2 Main Features of the Professional Blow Dryers

The chance to radically rethink the equipment used in the salons, given the consolidated product features and a competition based on quality/cost trade-off, leads to follow the branch of IPPR methodology treating the third class of business problems. The starting point of this activity has taken into consideration the redesign efforts dedicated to propose a new profile for a fundamental product in the hairdressing industry, i.e. the professional blow dryer. As recalled in Sect. 6.1, the most conspicuous innovation concerning such apparatus dates back to several decades ago. The development of ionic blow dryers has resulted in enhanced efficiency, without however considerably impacting hairdressers practices.

The introduction of small-sized hand-held hair dryers has fostered the diffusion of items for domestic use, which apparently present slight differences with respect to the devices employed in beauty salons. The key features of professional blow dryers stand in greater power and in the capability to perform their functions for longer times without overheating. However, given the maturity of the product technology and the presence of a wider marketplace, also the instruments for self drying at home show good, well-established and stable performances. Such items, on the basis of the consolidated main features, fit the application of methods for the measure and the maximization of customer satisfaction or any engineering model dealing with product platforms undergoing limited changes [5–7].

Hence, the collection of the fundamental features characterizing the hair dryers and markedly the devices for professional use, results in a quite easy task. An initial information gathering has been performed to obtain an overview about the most influential competing factors. It has involved the consultation of technical material available on the web and of some stylists, beyond the mentioned scientific sources. The task has resulted in the individuation of a prior set of product attributes, on which the sources have approximately converged and that can be summarized in the followings:

- design and esthetical qualities;
- cost;
- energy efficiency;
- ergonomic grip;
- peak power to speed up the hair drying;
- versatility of the temperature of the air jets;
- versatility of air speed;
- durability;
- ease of handling;
- ease of maintaining;
- ease of repairing;
- stability of performances during the use;
- avoidance of vibrations and noise;
- strength against shocks.

The investigation of the working conditions for the specific product can however support the elucidation of further (tacit or unfulfilled) customer requirements, as supported by [8].

6.3 Creating New Value Profiles Through IPPR

The present Section depicts the implementation of the tools and the procedure foreseen for the reengineering problems falling into class 3. The following subsections are dedicated to describe the sequence of activities involved in order to carry out the task, tackling the case study concerning the hair dryers employed by professional stylists.

6.3.1 Product Information Elicitation and Modeling for a Professional Blow Dryer

As Chap. 3.2.2.2 clarifies, the aim of the first activity concerns the collection of possible sources of value to be exploited in order to generate satisfaction for the buyer. In this case the customer is represented by the hair stylist or by the salon, purchasing the blow dryer for common working activities. The hairdresser represents plainly the end user of the product and thus the one directly perceiving the advantages ensuing from the usage of the device. According to the specific value chain, the benefits arising by the product usage can however involve, indirectly, the salon as a whole or its clientele.

The primary objective is thus the creation of a comprehensive *Lifecycle System Operator*. Such tool is intended to monitor the working conditions of the blow dryer, its relationships with further systems or subjects, the activities that precede or follow the display of the apparatus functions, regarding both the status of the product and the events occurring in the hairdressing shop. The results of the mapping procedure have been obtained throughout the submission of the tailored questions (as illustrated in Chap. 3.2.2.2) for the elicitation of valuable design inputs according to time and hierarchy dimensions. The queries have been administered to three volunteer respondents involved in the hairdressing field and the answers have been consequently joined in the framework of the Lifecycle System Operator, leading to the generation of the scheme reported in Table 6.1.

In order to clarify the use of the questions in the given context, we consider the environment in which the product is situated during the utilization time. The volunteers were asked to answer to the following question:

Are there any circumstances occurring during the <utilization time of the hair dryer> and concerning the <environment in which it is situated>, m to be observed and treated, potentially resulting as inputs for a valuable design of the product?

The answers of the consulted stylists have been collected and grouped, so to build a set of potential value sources, as in the followings:

- the suitability of the hair dryer according to the requested details for the requested hairdo, according to customer tastes and wishes;
- the presence of vibrations and noise produced by the hair dryer, which affect the surrounding environment;
- the presence of other systems within the salon such as chairs, the capes, the mirrors and, more in general, the furniture and other accessories;
- the presence of other customers within the saloon;
- the hot air produced by the hair dryer mistakenly directed to not haired surfaces of the customer body.

The collected sources of value have been then used to elicit the product attributes, which can plausibly contribute to determine the customer appreciation. The emerging sample of attributes has to be integrated with the already present (and overlapping) set of established features, that have been listed in Sect. 6.2, as a direct result of the preliminary product investigation, which were aimed at gathering the prior information about the blow dryers.

The questioning techniques proposed in Chap. 3.2.3.3 allow furthermore to determine the appropriate functional features related to the set of attributes. Additionally, throughout the gathering of the volunteers' opinion, an evaluation has been provided about performance levels and their correspondence with actual customer demands. To the purpose of clarifying the way of performing the classification of the attributes, let's take into consideration the following example. If the user of IPPR wants to classify the CR1—"Design and esthetical qualities", he/she must ask himself/herself and the stylists:

Table 6.1 Lifecycle system operator for a professional blow dryer

	Purchasing, choice and access activities	Before use operations	Utilization time	Elapsed time before further exploitations	End of the functioning
Environment in which the product is situated	Service for transportation; Installation; Training	Stylist; Decision about the hairdo; Salon, chairs, capes and mirrors; Other customers being served or waiting	Use suitable for the stylist; Vibrations and noise; Salon, chairs, capes and mirrors; Customer (and further details of the hairdo); Other customers being served or waiting to be served (sitting, reading magazines, talking); Very hot jets of air towards not haired surfaces of the customer	Customer check about the result (i.e. mirroring the nape or the back); Cleaning floor; Chairs, capes, mirrors and towels; Hair health; Stylist health	Salon renovating
Product or service level	Cost; Efficiency, durability and main features; Design and esthetical qualities	Hair washing; Hair cutting; Hair dying	Hair drying and styling (with pauses for hair brushing); Combined use with hairbrush; Controllability and versatility of the hair drying (amount of water, hair length, hairstyle, hair color, kind of hair); Use of energy; Blow dryer wire and connection to a fixed surface; Alternative employment of straighteners, curling wands	Storing; Tidying up; Reducing temperature; Maintaining; Repairing	Disposing; Recycling; Reusing in other environments; Reusing with further damaged parts
Parts, components and accessories	Components		Components during use	Components after use; Refilling	Disposed materials

if we consider the <Design and esthetical qualities>, are we dealing with the endeavor to request the customer less money, time, energy, space, tools, materials, information, experience or know-how?

The aesthetic characteristics are properties which have not any impact on the resources consumption of the customer, therefore the answer is negative. Hence, the following question arises, as suggested by the procedure:

if we consider the <Design and esthetical qualities>, are we dealing with the objective of reducing the impact of an undesired event, generally associated with the product functioning or decrementing the probability of such unwanted situation?

Once again the answer is negative since the considered feature is not related to the mitigation or elimination of damages or bad consequences that the customer can undergo, in a direct or indirect way, due to the hair dryer. Thus, the user must answer to the last question:

if we consider the <Design and esthetical qualities>, are we dealing with the effort of increasing the benefits for the customer or for a circumscribed group of users, the versatility of the product functioning, the stability of the outcomes, the delight generated by the treated system?

A pleasant aspect of the product is a property which aims at delighting the user. These considerations allow to answer affirmatively the question, thus the "Design and esthetical qualities" can be suitably classified as an attribute characterized by the "UF" functional features.

Further on, the analysis of the present hair dryers reveals a certain care towards the aesthetics; however, with regards to the fashion environment in which the devices are situated enhancements would be welcomed. Therefore, the offering level of this attribute can be assessed as moderate and barely sufficient to satisfy the present demand of the customer.

Since the value profiles of professional blow dryers do not differ considerably, the framework reported in Table 6.2 has been deemed sufficient to map the main aspects about the market of the devices under investigation. The Table reports the monitored product attributes, whereas the first 14 items are directly related to the competing factors listed in Sect. 6.2, while the others have been obtained by focusing the attention on the value sources mapped through the Lifecycle System Operator.

The procedure allows to point out, beyond the functional features relevant for the fulfillment of the next task, those attributes which result missing or unsatisfied, viable to indicate interesting business opportunities regardless the employment of IPPR suggested tools. Besides, the big quantity of product attributes, as listed in Table 6.2, shows a noticeable amount of differentiation opportunities for the investigated industry. Thus, several new product profiles can be generated through the employment of the suggested guidelines.

Table 6.2 Product attributes emerging from the product analysis

CR#	Product attribute	Functional feature	Performance and customer demand
1	Design and esthetical qualities	UF	Moderate, barely sufficient
2	Cheapness of the device	RES	Good, adequate
3	Energy efficiency	RES	Moderate, unsatisfying
4	Maneuverability, due to ergonomic grip	RES	Good, adequate
5	Hair drying speed (due to peak power)	RES	Good, often outstripping the demand
6	Versatility of the temperature of the air jets	UF	Very high, outstripping the demand
7	Versatility of air speed	UF	Good, adequate
8	Durability of the device (along the time)	UF	Good, barely sufficient
9	Avoidance of slipping from the hands	RES	Moderate, unsatisfying
10	Ease of maintaining	RES	Low, unsatisfying
11	Ease of repairing	RES	Low, unsatisfying
12	Stability of the performances during the use	UF	Very high, adequate
13	Avoidance of vibrations and noise	HF	Low, unsatisfying
14	Strength against shocks	HF	Moderate, adequate
15	Ease of installation	RES	Very high, adequate
16	Limitation of needed training	RES	Very high, adequate
17	Provided support along the operations preceding the styling	UF	Absent, unsatisfying
18	Reduced encumbrance	RES	Moderate, unsatisfying
19	Avoidance of overheating during use	HF	Good, barely sufficient
20	Controllability of the air jet direction	UF	Good, barely sufficient
21	Avoidance of heating not haired human skin	HF	Absent, unsatisfying
22	Limitation of noise, silentness	HF	Moderate, unsatisfying
23	Limitation of vibrations	HF	Good, adequate
24	Versatility in accomplishing variable customer requests about the hairdo	UF	Moderate, unsatisfying
25	Versatility in treating various kinds of hair	UF	Low, unsatisfying
26	Possibility to be replaced with alternative instruments	UF	High, adequate
27	Freedom of use, maneuverability (due to external boundaries, i.e. electrical wire, hairbrush)	RES	Low, unsatisfying
28	Aesthetical matching with salon equipment	UF	Absent, unsatisfying
29	Functional matching with salon equipment	UF	Absent, unsatisfying

(continued)

Table 6.2 (continued)

CR#	Product attribute	Functional feature	Performance and customer demand
30	Controllability of the on/off functions	UF	Good, barely sufficient
31	Limitation of energy consumption due to unintended use	RES	Low, unsatisfying
32	Customizability	UF	Absent, unsatisfying
33	Versatility of the device according to stylist's experience and individual characteristics	UF	Low, unsatisfying
34	Possibility of integrating accessories	UF	Moderate, unsatisfying
35	Possibility of replacing damaged components	UF	Moderate, unsatisfying
36	Ease of replacing damaged components	RES	Low, unsatisfying
37	Limitation of the probability of experiencing damages along pauses of use	HF	Good, adequate
38	Limitation of hurdling activities in the salon along pauses of use	HF	Good, adequate
39	Ease of maintenance	RES	Good, adequate
40	Cheapness of maintenance	RES	Low, unsatisfying
41	Safety (e.g. limitation of carpal tunnel)	HF	Low, unsatisfying
42	Capability to blow away cut hair	UF	Good, adequate
43	Possibility to be reused after the failure	UF	Low, unsatisfying
44	Reusability of the components	UF	Absent, unsatisfying
45	Recyclability	UF	Moderate, barely sufficient
46	Reusability in event of salon changing or renovating	UF	High, adequate
47	Environmental sustainability	HF	Moderate, barely sufficient

6.3.2 Building a New Profile and a Preliminary Conceptual Idea for a Professional Blow Dryer

The present Section shows the application of the proposed IPPR tools to generate a feasible original profile for the investigated case study. According to the iterative redesign process illustrated in Chap. 3.3.2, the task has to be fulfilled by alternatively employing methods and techniques relevant for both the steps 2 and 3 of the overall methodology, as structured for the third class of reengineering problems.

By following the footprints of the recalled cyclical roadmap, the first recommended stage is the individuation of priority objectives to be pursued through the product rethinking. In coherence with the indications provided by the New Value Proposition Guidelines (NVPGs) and with the need to act on insufficient performances, the main endeavor has been addressed towards the attainment of a more practical employment of the blow dryer (improvements about the CR27, classified as RES) and reduction of health related side effects (enhancements regarding the CR41, clustered as HF). The goals are somehow related, since an increased ease of handling could allow to reduce drawbacks concerning the emergence of muscular problems on stylists, such as the carpal tunnel syndrome, to which the weight of the blow dryer contributes. The twofold scope can be accomplished by introducing solutions that do not require the front wire, a relevant hurdle for gaining a suitable maneuverability, and characterized by a low weight of the apparatus.

The presence of a limited bundle of attributes to be analyzed has allowed the immediate identification of the related engineering requirements and design choices. The employment of the Quality Function Deployment, suggested in the workflow of the methodology, can be skipped with regards to the presented case study.

Subsequently IPPR recommends to delineate how to implement the previous advantages. In order to increment the lightness of the device, a traditional design approach would head towards a trade-off between the weight of the engine and the power of the apparatus. The latter engineering feature determines the speed of drying (CR5, RES), a sometimes oversupplied characteristic that is however better not to be jeopardized, especially because of the indications emerging in the NVPGs. As an alternative, cordless battery blow dryers could be proposed, but their performances do not fit the exigencies of hairdressing applications.

The overcoming of the contradiction between mutually not compatible demands (lightness and power) requires proper techniques leading to valuable conceptual solutions. In such situations, TRIZ, as highlighted in Chap. 3.4.2, includes the most helpful body of knowledge.

For instance, by exploiting inventive and/or separation principles, the resolution of the conflict heads to concepts aimed at disconnecting the components of the hair dryer. Indeed, the element performing the main useful function of the apparatus, i.e. the tube conveying and directing a hot flow, can be separated by the fan addressed at providing the requested properties of the air (by heating and accelerating it). The transformation pattern can be supported by rearranging the collocation of the device (and of its parts) inside the salon. The displacement of the blow dryer motor into elements belonging to the salon environment has been thus evaluated, individuating two possible solution concepts. The options regard the integration of a body including the engine within the roof or the salon chair, with the consequent handling of a flexible tube to be easily moved and directed by the stylist. The ideas could be roughly depicted by positioning a hairdryer suite, like those commonly employed in the locker rooms of sports facilities, in the intended points of the shop, like sketched in Figs. 6.1 (for the roof) and 6.2 (for the salon chair). The rough concepts require undoubtedly a proper sizing of the tube length and a suitable design of the sensing mechanism capable to activate the requested flowing of hot air. Nevertheless no sophisticated technology is expected to be required in order to implement the presented ideas.

At a first glance, the thought solutions could be relevant for all the stylists, given the diffusion of muscular diseases due to the employment of hairdressing tools. Moreover, the elimination of the front wire can represent a benefit per se, given the consequent nuisance arising as the dryer is handled and sometimes moved in front of the customer. However it is deemed that less powerful hairdresser could perceive the greatest advantages.

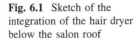

Fig. 6.1 Sketch of the integration of the hair dryer below the salon roof

The following step to be run regards the identification of possible additional benefits or potential drawbacks, emerging as the development is carried out of the ideas kept in the "rough copy" of the project.

In the first case (roof implementation of an engine feeding one or a plurality of blow dryers in the salon), the distance of the motor would result in a diminished noise, if the area occupied by the stylists and the customers is considered. The decrement of noise, thus the application of the Raise action for the CR22, would result in a further considerable advantage, according to the NVPGs and the general criteria concerning the employment of the FAF. On the other hand, in the second option (salon chairs equipped by a component, resembling a typical wall hairdryer suite), the presence of the engine in the surroundings of the working space would not considerably impact the silentness of the environment. Thus, in order to provide a major noise restraint, the peak power of the blow dryer should be significantly limited. This means to resort therefore to the action Reduce CR5 (RES), about which we have already warned out, due to the infringement of the NPVGs.

In both the cases, the presence of flexible tubes favors the introduction of a control mechanism meant to automatically and quickly stop the air jets when unneeded. Such control could be operated by the stylist with simpler movements of the hands and the fingers, e.g. by releasing the tube. This measure would positively impact the controllability of the air jet, leading to the action Raise CR30 (UF), moderately impacting the success of a NVP, according to the guidelines.

The foreshadowed pair of conceptual solutions would allow even an integrated functional design of a set of salon equipment and accessories. The consequent action, Create CR28 (UF) is complaint with the advantageous measures foreseen within the grid of NVPGs.

The main drawback of the design concepts could result in a more difficult regulation of the temperature ranges, whereas such function is allowed by directly operating the buttons on common blow dryers. The resulting measure (Reduce CR6, UF) is allowed by the general rules for the application of the FAF and is meant to cause limited bad consequences with respect to the guidelines.

Furthermore, the possibility to use special hair drying and styling tools, that sometimes replace the traditional hair dryers (e.g. straighteners, curling wands), would be completely jeopardized, if not through their common connection to the

Fig. 6.2 Sketch of the
implementation of the hair
dryer within the salon chair

plug. The additional action to be considered (Eliminate CR26, UF) is not conso-
nant with the scope of the FAF, being such attribute not oversupplied, but does not
represent a particular source of harm with regards to the proposed guidelines.

Other features could be automatically impacted by the design of the novel
product profile, but at a marginal level. For example, the possibility to easily
switch off the hair dryer could bring improvements in terms of reducing energy
consumption, avoiding to generate hot air in the short periods it is not needed and
the device is not turned off for the sake of convenience, e.g. when the hairdresser
moves around the customer to be styled. Additionally the same conditions could
result in negligibly less probable events of directing hot air towards the skin of the
salon visitor. In absence of consistent value shifts according to these features, no
action is considered highlighting such advantages.

On the basis of the followed redesign procedure, Table 6.3 summarizes the actions
that would result in the NVP task, whereas the fifth one, reported in italics, concerns just
the solution involving the integration of the blow dryer and the salon chair. According
to the depicted transformation of the product profile, the number of value-adding
measures is higher than the occurrences of Reduce and Eliminate actions, especially for
the roof solution. At the same time the infringement of the guidelines involves just
partially the option of the joining of the treated device with the chair. The team
formulating the present NVP strategy has deemed the results satisfying for the verification and the consequent further development of the profile. Therefore, no additional
cycle of the roadmap procedure described in Chap. 3.3.2 has been performed.

6.4 Survey and Discussion of the Results

In order to assess the appreciation of professionals about the new generated profiles, and potentially the differences between the alternatives, a suitable ques-
tionnaire was submitted by email to some randomly identified hairdressing shops
and beauty salons. The authors received 30 replies, among which some of them
resulted incomplete. The main outcomes can be summarized as follows:

Table 6.3 New value propositions for a professional blow dryer

Action	Functional feature	Product attribute
1. Create	UF	CR28 Aesthetical matching with salon equipment
2. Raise	UF	CR30 Controllability of the on/off functions
3. Raise	HF	CR22 Limitation of noise, silentness
4. Raise	RES	CR27 Freedom of use, maneuverability
		(due to external boundaries, i.e. electrical wire, hairbrush)
5. Reduce	RES	CR5 Hair drying speed (due to peak power)
6. Reduce	UF	CR6 Versatility of the temperature of the air jets
7. Eliminate	UF	CR26 Possibility to be replaced with alternative instruments

- the enhancements with respect to common blow dryers result effectively as a considerable benefit according to the opinion of 27 respondents out of 29;
- 22 stylists out of 26 would take into account the advantages shown by the presented hypothetical products, when choosing the equipment and the furniture for a new or renovated salon;
- according to 20 answers out of 26, it is affirmed that similar products could be likely to be launched in the marketplace in the next future;
- however, just 12 hairdressers out of 26 would tolerate the drop of the hair dryer performances in terms of peak power.

The results of the test confirm the attractiveness of value profiles that follow the guidelines emerged by the presented research. On the contrary, the infringement of some of them, such as the increase of resources consumption in terms of time within the analyzed case study, results in consistent drops in customers appraisal. In this sense, the solution concerning the displacement of the hair dryer body below the roof, results more convincing.

Insights about the motivations of the respondents, with reference to the acceptability of the reduction of the peak power, give rise to different ideas and approaches in making the business. The stylists that would easily bear such diminishment, state that the disproportionate power of the blow dryers results in exceeding temperatures, leading to ruin the hair. On the other hand, the drying speed does not result a relevant competing factors for those kinds of salons, welcoming a longer staying of the clients in the shop. The latter regards mostly small hairdressing outlets with a loyal and consolidated clientele, for which the visit to the stylist involves also spending good time and keeping social relationships. Similarly the need to serve quickly the customers is not perceived by the hairdressing shops including additional beauty treatments and spa. In such cases the long duration of the visits results as a greater arouse for the customers of the salon and an occasion for the shop holder to offer additional services. Conversely, the outlets serving customers with low amounts of available time for hair care, cannot disregard the drying power of the device and the speed in performing the styling.

As a result, the new product profile and its alternative implementations underline a segmentation of the current marketplace, according to circumstances that have not been yet fully investigated by the hairdressing industry. Such indications are widely confirmed by researches about customer loyalty, according to which age and social factors play a fundamental role in the client retention of hairdressing stores [9, 10].

Thus, the task of tackling a reengineering problem of the third class, has resulted in the individuation of a favorable direction and inspiration for the innovation of the equipment used in hairdressing salons, beyond the generation of a new product profile, worth to be further developed. In such framework we refer particularly to the solution involving the integration of the hair dryer in the surroundings of the shop roof, which best corresponds to the indications addressed by the NVPGs. Within the whole NPD cycle, existing patents represent a valuable starting point for the generation of a physical embodiment of the innovative blow dryer, e.g. [11, 12], potentially leading to provide the technical support for an enhanced business model.

References

1. White G, Croucher R (2007) Awareness of the minimum wage in the hairdressing industry: research report for the Low Pay Commission, UK
2. Lee T, Ashton D, Bishop D, Felstead A, Fuller A, Jewson N, Unwin L (2005) Cutting it: learning and work performance in hairdressing salons. In: 4th international conference on researching work and learning on challenges for integrating work and learning, Sydney, 11–14 December 2005
3. Ryan C, Watson L (2003) Skills at work: lifelong learning and changes in the labour market. Research for the Australian government, department of education, science and training, ISBN 0-642-77390-4
4. Habia (2006) Skills foresight for the hair and beauty sector 2006, UK
5. Leung P, Ishii K, Abell J, Benson J (2008) Distributed system development risk analysis. J Mech Des Trans ASME 130(5):051403 (9 pages)
6. Chen WL, Chiang YM (2010) A study on the product design of hair dryer using neural network method. In: 2010 international symposium on computer, communication, control and automation, Tainan, 5–7 May 2010
7. Olewnik A, Hariharan VG (2010) Conjoint-HoQ: evolving a methodology to map market needs to product profiles. Int J Prod Dev 10(4):338–368
8. Kain A, Kirschner R, Gorbea C, Kain T, Gunkel J, Klendauer R, Lindemann U (2010) An approach to discover innovation potential by means of DELTA applications. In: Design conference 2010, Dubrovnik, 17–20 May 2010
9. Patterson PG (2007) Demographic correlates of loyalty in a service context. J Serv Mark 21(2):112–121
10. Chen SC, Quester PG (2006) Modeling store loyalty: perceived value in market orientation practice. J Serv Mark 20(3):188–198
11. Hopper MB (2006) Tool support. US Patent 7150424
12. Bazzicalupo LM, Mangiarotti R (2008) Hairdryer device. EP Patent 1973442

Chapter 7
Discussion and Concluding Remarks

7.1 Introduction

This Chapter is devoted to an overall discussion about IPPR with the aim of summarizing its main features, weak points which require to be overcome and results achieved during the experimentation activities.

More in particular, the purpose of Sect. 7.2 is to provide an overview of the overall achievements with respect to the methodological objectives presented in Chap. 1. With the aim of shedding light on the current priorities to attain methodological and practical improvements, the research activities carried out to develop IPPR are synthetically illustrated.

Section 7.3 briefly summarizes the results of the performed tests, focusing on the evaluated performances of the method in terms of effectiveness, robustness and repeatability, thus providing some suggestions to enhance the methodology.

7.2 IPPR: Achievements and Open Issues

The achievement of the proposed methodological objectives is hereby discussed with the aim of pointing out the attainments as well as the issues still open, which require further research efforts.

Altogether, IPPR is a decision support instrument which provides a viable aid in identifying the most appropriate approach to solve the addressed reengineering problem. The outputs supplied by the method consist in a clear and comprehensive representation of the priorities and the consequent redesign actions to be undertaken. As stated in Chap. 1, successful BPR initiatives require an exhaustive description of the process functioning in order to highlight the main deficiencies on which to concentrate the reengineering efforts. This task can be performed by using modeling techniques capable to summarize the whole set of information and

F. Rotini et al., *Re-engineering of Products and Processes*,
Springer Series in Advanced Manufacturing, DOI: 10.1007/978-1-4471-4017-7_7,
© Springer-Verlag London 2012

data belonging to different domains. To this end, IPPR provides a process modeling technique which allows the collection and structuring of all the information and data related to the business process, in both technical and economical domains. The proposed model is based on the integration of different techniques such as:

- IDEF0 to represent the process activities according to the scheduled sequence, indicating involved know-how, employed technologies, control mechanisms;
- EMS to account for the flows of energy, materials and information involved in the business process;
- TOC, to represent the expenditures related to each phase of the business process.

Beyond the process modeling activity, the determination of the relevant aspects of value delivered to the customer, represents another crucial task in order to obtain meaningful feedbacks about what should be changed in the business process. Reasoning tools have been developed within IPPR to support this step, such as the Lifecycle System Operator and the CRs Checklist. These instruments are suitable means which allow a comprehensive investigation of the aspects of value ascribable to a given product. They characterize the product along the dimensions of its lifecycle and according to the functional role played by its features within the delivery of customer satisfaction.

A further original contribution of IPPR regards the extension of the Value Engineering validity beyond the classical approach which assesses the worthiness of a process through the ratio between technical performances and involved costs. Indeed, proper assessment metrics have been defined that consider the rate between the contribution of each phase in generating the customer satisfaction and the resources required to deliver such benefits.

The reengineering directions arising by the employment of IPPR for problems concerning industrial processes are oriented towards the growth of the value associated with the delivered outputs and/or with the reallocation of the resources along the sequence of the phases.

Furthermore, suitable guidelines have been defined to support new value proposition tasks oriented towards the attainment of radical innovations for products and services. Such guidelines have been extrapolated thanks to a deep analysis of dozens of success stories and represent a complementary tool to strengthen the application of the Four Action Framework.

The developed tools to guide the redesign of product platforms is a consistent aid also within the field of Product Service Systems. The generation of new product profiles can include the introduction of a major servicing, thus disclosing when such measure could result suitable for providing enhanced customer value.

Despite the consistency of the above summarized achievements, IPPR requires further developments, aimed at primarily improving its systematic degree. Some of these issues are already addressed in ongoing research activities.

Among the various criticalities, the identification of the right level of detail at which to deepen the modeling activities is still an essential issue to be addressed.

With reference to process reengineering, the quantity of the phases to be investigated can influence the identification of the relevant process bottlenecks, which represents a key step in IPPR. Therefore, the definition of more precise rules aimed at supporting the user in the segmentation of the process into a meaningful number of phases according to the customer requirements to be fulfilled, is an essential development activity in order to improve the reliability of IPPR.

As widely underlined in the book, IPPR employs the value indexes as fundamental criteria for highlighting the process criticalities. Such coefficients result as a combination of factors concerning the expected delivered benefits and the costs addressed at the industry level. The evaluation of the bottlenecks through these metrics is still performed regardless of the impact on customer satisfaction resulting by fluctuations of the actual process efficiency. The latter influences indeed the offering levels pertaining the fulfilled product attributes. Thus, the integration is strongly recommended of the developed metrics with models capable to take into account also the recalled efficiency aspects, in order to strengthen the rigor of IPPR as decision support system.

Further on, with a particular emphasis on the information gathering, methodological deficiencies arise due to the disproportioned role entrusted to the knowledge of the sector experts for a wide range of analysis activities. Such tasks include the definition of the customer requirements, their relevance in determining the customer perceived value and their classification according to the Kano model. The reliability of the outputs provided by these activities can benefit from a more systematic approach in order to reduce the impact of subjective evaluations. Besides, the introduction of a model capable to manage diverging opinions about the issues investigated along the analysis of the business process, could definitely make these activities more robust.

Eventually, other research activities have been outlined to improve the potential of the guidelines for supporting NVP tasks. In such a context, the definition of a more prescriptive path as well as the enhancement of the classification of the product features, are essential activities to strengthen the application of the guidelines. A planned activity regards the fine-tune of a metric to assess the likelihood of product success, according to the followed value transition. Such an instrument, currently in its validation phase, could operate as a decision support about investments to be channeled in light of a plurality of product platforms alternatives to develop within a company.

According to the above discussion, it can be concluded that the methodological objectives have been overall attained. However, some developed tools and some application steps require further investigation in order to enhance the effectiveness of IPPR.

7.3 Reliability of IPPR

The book illustrates applications of the proposed tools to three case studies, with regards to the different classes of business problems described in Chap. 1. Beyond the step-by-step application of the methodology, Chaps. 4–6 include a final discussion enforcing the conclusions emerging as a result of IPPR employment, throughout surveys and crosschecks using literature sources. Hereby, we briefly recall the issues supporting the outcomes of the methodology.

With reference to class 1 of reengineering problems, the logic and the tools for process analysis have been proposed in the field of the solid bio-fuel production process. Starting from the analysis of the market opportunities and process needs, the application of the method has brought to identify the need of designing an innovative dewatering and grinding technology for woody biomass capable to improve the efficiency of the whole manufacturing process. The arisen indications have been widely verified through the well established and acknowledged scientific literature.

According to the second class of business problems, the method has been applied in the footwear sector bringing to the definition of suitable reengineering directions aimed at shortening the time to market in the industry of accessible fashion shoes. All the identified strategic actions comply with those successfully implemented in the sector and arisen from dedicated research projects.

With regards to class 3, the guidelines for New Value Proposition have been applied in the field of the professional hair dryers. IPPR supported the definition of new relevant product features, starting from the analysis of the customer needs and the survey of the devices currently in the marketplace. The effectiveness of the arisen indications has been verified by using the Voice of the Customer, as a means to obtain feedbacks in the prior product development stage. The results have shown that a new value profile compliant with the guidelines, meets the consensus of a not negligible segment of users (about 90%).

Beyond the application of the presented tools in different industrial domains, which are not reported in the book, tailored activities have been carried out with the aim of verifying the robustness and the repeatability of outcomes of the method. Within process reengineering tasks, a case study belonging to the pharmaceutical field has been considered for an experiment, involving a sample of MS engineering students. The example is related to established industrial practices which have been overcome and partially substituted by known process alternatives. Even if the size of the testers group was not sufficient to perform a fully acceptable validation, some interesting evidences arose. The test revealed an overall consistency with the results extracted by sector experts employing IPPR and a fundamental coherence with the redesign of the industrial activities observed in the technical evolution of the process.

Although a great amount of individual tests converge towards the results extracted by the sector experts, the sample of students provided not negligible variable outcomes, potentially leading to misleading conclusions. With respect to such condition, it is recommended to perform multiple applications of the method by different users in order to take into account a plurality of points of view within the company or the industry and to collect the gathered data. By applying statistical tools treating the diverging inputs, the end results are characterized by enhanced reliability. In order to ease the task, the authors are developing a computer aided tool to gather the evaluations of multiple analyzers and extract the most consistent process bottlenecks.

The tools for product reengineering have been subjected to wide experiments, involving academics and University students. Whereas the possibility to generate innovative product profiles still depends on individual creativity, the proposed tools have demonstrated a consistent capability in guiding the exploration of potential value adding circumstances. This results in a severely augmented capability to individuate valuable alternatives to undergo product development initiatives. Additionally, a pair of PhD students is currently in charge of the methodological development of the New Proposition Guidelines, with the objective of delivering a computerized decision support system assisting the early stages of NPD tasks.

7.4 Final Considerations

Eventually, the manifold applications of the methodology in very different contexts have demonstrated its effectiveness in identifying value bottlenecks and suitable directions for overcoming the main hurdles. The original contribution of the book, although the presented methodology will be subjected to continuous improvements, is a set of (already effective) practical suggestions to orientate reengineering initiatives in industry, instead of relying on the most convincing techniques proposed by consultants which cannot fit all the situations. The changing conditions in the marketplace and at the industry level require indeed customized tools; let's attempt to aid companies at least to choose the most suitable ones. The change is the motor of innovation, perhaps also the greatest source of risks and opportunities for the enterprises. For sure, it is the main fuel of our research activities.

Appendix A
The IDEF0 Model

The set of models belonging to the Integration Definition Function Modeling (IDEF) group refer to a family of standards for systems and functions representation developed under the aegis of the United States Air Force.

IDEF0 represents one of the most diffused models within the group of IDEF schemes and with reference to the set of methods employed to depict decisions, actions and activities. By providing the functional perspective of a given system, the representation scheme is aimed at facilitating the communication among different parties. Beyond supporting the interchange of information, the model helps the analyst in focusing on which functions are delivered by the system and what enables its working mechanism.

According to what is claimed in the registration document of the model at the American National Institute of Standards and Technology (NIST), the functional language owns the following features and capabilities:

- "Performing systems analysis and design at all levels, for systems composed of people, machines, materials, computers and information of all varieties—the entire enterprise, a system, or a subject area;
- Producing reference documentation concurrent with development to serve as a basis for integrating new systems or improving existing systems;
- Communicating among analysts, designers, users, and managers;
- allowing coalition team consensus to be achieved by shared understanding;
- Managing large and complex projects using qualitative measures of progress;
- Providing a reference architecture for enterprise analysis, information engineering and resource management" [1].

According to its formalism, IDEF0 models illustrate the relationship of all the functions in a graphical format with "box and arrows", whereas the boxes are the functions themselves and the arrows stand for the constraints and the involved flows.

F. Rotini et al., *Re-engineering of Products and Processes*,
Springer Series in Advanced Manufacturing, DOI: 10.1007/978-1-4471-4017-7,
© Springer-Verlag London 2012

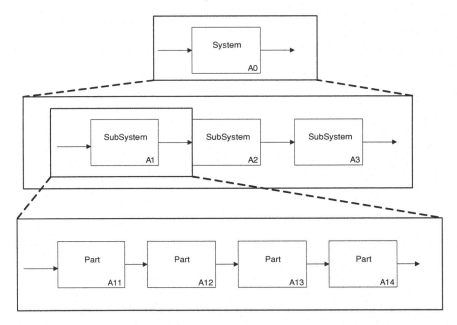

Fig. A.1 Hierarchies of systems and functions in the IDEF0 model

The description starts with a single box, standing for the representation of the overall designed system, labeled as A0. The box A0 is subsequently decomposed into a more complex diagram constituted by up to seven interconnected boxes. The hierarchical subdivision is repeated for each function of the diagram, then for each box in the resultant schemes and so on, until the system is fully described. Boxes are numbered according to the hierarchical levels, with the label being used to facilitate the comprehension of the relationships among different diagrams. For instance, the first box resulting by splitting the A0 is A1, and the third box in the decomposition of the latter is A13 [2]. Figure A.1 clarifies the decomposition process.

The flows depicted through the arrows represent manifold sorts of inputs, outputs and controlling rules. The inputs in form of materials or objects to be transformed enter the box from the left; the arrows directed towards the top of the box refer to controls, whereas those reaching the opposite side represent mechanisms and technologies. The outputs of the system exit the box from the right side (see Fig. A.2).

Although IDEF0 standard has been conceived as a functional modeling tool, it is often used even to represent processes, since the system functions that can be modeled include activities, actions, processes, operations [1].

As a result, within process modeling, IDEF0 models can be employed with a different formalization: the boxes describe the activities [3] that perform the functions and the arrows stand for the information and the objects that are interrelated in a given system [4]. Due to such common application, IDEF0 is

Fig. A.2 Inputs and outputs
in the IDEF0 model

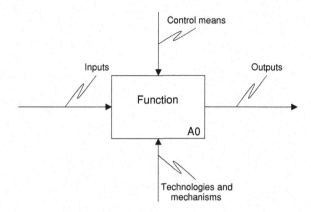

considered among the most useful instruments also to perform business process modeling, with the possibility to decompose the processes in lower level activities [5].

Appendix B
The EMS Model

The Energy-Material-Signal (EMS) model is a sort of black box representation to schematize functions, based mainly upon the work by Pahl and Beitz [6] and other scholars. An analysis of engineering systems reveals that they essentially channel or convert energy, material or signals to achieve a desired outcome. Energy is manifested in various forms including optical, nuclear, mechanical, electrical, etc. Materials represent matter or substances. Signals represent the physical form in which information flows. For instance, data stored on a hard drive (information) would be conveyed to the computer's processor via an electrical signal. An engineering system can therefore be initially modeled as a black-box (Fig. B.1) with energy, material and signal inputs which are modified from the system in the form of outputs. According to EMS original formalism, energy is represented by a thin line, material flows by a thick line, and signals by dotted lines as shown. The engineering system therefore provides the functional relationship between the inputs and the outputs [7]. The system can be further subdivided into sub-systems of a lower hierarchical level to better describe the involved functions and transformations.

Fig. B.1 Sketch of the EMS model

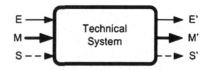

F. Rotini et al., *Re-engineering of Products and Processes*,
Springer Series in Advanced Manufacturing, DOI: 10.1007/978-1-4471-4017-7,
© Springer-Verlag London 2012

Appendix C
The Model of the Theory
of Constraints (TOC)

The Theory of Constraints [8] is a management practice aimed at identifying the weak ring of a value chain, according to financial metrics.

According to the TOC, the production process is represented as a technical system constituted by chains of operations, where each ring represents a phase; in Fig. C.1 an example is provided. The flows taken into account in this kind of model are the monetary flows generated by the system that are defined as it follows [9]:

- Throughput (T): "The rate at which the entire system generates value through sales (product or service)": this flow represents the money coming in the system.
- Inventory (I): "All the money the system invests in things it intends to sell": this is the flow of money that is spent in order to buy raw materials.
- Operating Expenses (OE): "All the money the system spends turning Inventory into Throughput", this flow of money going out the system to buy labor, utilities, consumable supplies, energy, etc.

With regards to the convention introduced by TOC, the throughput is the revenue coming from sales, divided by the time elapsed to carry out the manufacturing of the items.

F. Rotini et al., *Re-engineering of Products and Processes*,
Springer Series in Advanced Manufacturing, DOI: 10.1007/978-1-4471-4017-7,
© Springer-Verlag London 2012

Fig. C.1 Sketch of the activities involved in an industrial process depicted through TOC formalism

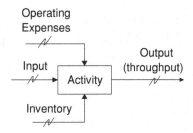

As a result of the individuation of the value bottleneck, in terms of throughput metrics, the endeavors of the firm should be channeled massively to improve the least performing stage of the business. Such measure should be ceased just when another segment of the industrial process becomes the new weak ring. Still according to TOC principles, the first priority in improving the system is to increase T, since it has the greatest potential impact on the bottom line, while decreasing OE and/or I is secondary and in any case it should not result in jeopardizing future throughput [10].

Appendix D
The System Operator

System Operator or, as Genrich Altshuller (the founder of TRIZ theory) named it, Multi-screen Schema of Powerful thinking, shows the model of advanced thinking in the course of problem solving process. Learning this model and developing appropriate skills to use it in practice is a core of Altshuller's educational program [11].

The System Operator can be used with different objectives within the problem solving process. For example, during the preliminary stages, while looking for roundabout problems whose solution allows to obtain the same overall goal, a multi-screen view helps orienting the thought from cause prevention to effects compensation or mitigation, as well as a means to change the scale of the solution space in order to avoid psychological inertia. Besides, while looking for resources, the System Operator helps focusing the attention on each relevant aspect of the system and its environment, by analyzing any time stage at any detail level with a systematic approach.

System Operator could be seen as a three-dimensional parametrical space (Fig. D.1):

- Dimension of Hierarchical level of System: Whatever is the element we are taking into account (System), it is always possible to consider its constituting parts (Subsystems) as well as the environment it belongs to (Supersystem),
- Dimension of Time: whatever is the time interval taken into consideration for a certain analysis or description (Present), it must be considered as a phase of a sequence, therefore with a Past and a Future;
- Dimension of Anti-Systems: whatever is the property of an element taken into consideration, this dimension suggests looking to the opposite values of the same property (anti-property); similarly, a combination of anti-properties characterizes an anti-system.

F. Rotini et al., *Re-engineering of Products and Processes*,
Springer Series in Advanced Manufacturing, DOI: 10.1007/978-1-4471-4017-7,
© Springer-Verlag London 2012

Fig. D.1 System operator or classical TRIZ multi-screen schema of powerful thinking [12]

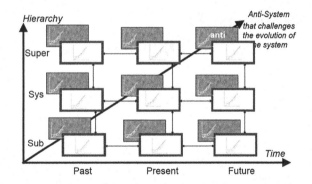

Disregarding its original formulation, several applications of the System Operator for the scope of browsing the properties of products or processes neglect the last dimension.

For practical needs it is useful to treat each of these three dimensions as a composition of several dimensions. For instance, in practice we often face with a situation that one Element belongs to several hierarchies of systems. An airbag in the car belongs to the dashboard or the doors or the steering wheel and, at the same time, it deals with the safety of the driver and the passengers.

Depending on the specific situation we can consider Time dimension as a historical time (if we study evolution of certain systems), as a process time (while analyzing a chain of events, even with their cause-effect relationships), as a life cycle of an element of a system.

Appendix E
The Kano Model of Customer Satisfaction

The Kano model [13] is a theory to support product development, whose main hypothesis stands in the diverse contribution of product features in impacting the perceived customer satisfaction. The model attempts to overcome a classical and tacit assumption, according to which the increased level of a product performance results in a proportional growth of value for customers.

The Kano Model classifies the product attributes on the basis of their effect on customer satisfaction. It divides the relevant attributes, generally defined as Customer Requirements (CRs) in three different categories that play a different role in the product or service perception: Must-Be, One-Dimensional and Attractive.

Must-Be CRs are attributes expected by the customer, they do not provide an opportunity for product differentiation, since they are commonly accomplished also by the competitors. Increasing the performance related to these attributes provides diminishing returns; however the absence or poor performance of these attributes results in extreme customer dissatisfaction.

One-Dimensional CRs are those whose performance growth results in linear enhancements of customer satisfaction. Besides, an absent or weak performance attribute determines customer dissatisfaction.

Attractive CRs are usually not explicit and unexpected by customers but can result in high levels of customer satisfaction, while their absence does not lead to dissatisfaction. These excitement attributes often satisfy latent needs customers are currently unaware of. In a competitive marketplace where manufacturers' products provide similar performances, obtaining excitement attributes that address "unknown needs" determines a substantial competitive advantage [14].

With regards to the presented logic, whereas Must-Be and Attractive attributes pertain just the avoidance of dissatisfaction and the generation of delight, respectively, One Dimensional CRs influence both the dimensions. Eventually,

F. Rotini et al., *Re-engineering of Products and Processes*,
Springer Series in Advanced Manufacturing, DOI: 10.1007/978-1-4471-4017-7,
© Springer-Verlag London 2012

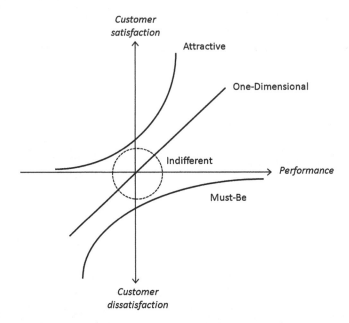

Fig. E.1 The Kano model of customer satisfaction, relating the performance of the fulfilled customer requirements and the impact on satisfaction/dissatisfaction according to the kind of product attributes

poorly relevant attributes, which can be easily disregarded are described as Indifferent attributes. The whole framework can be summarized by means of the diagram in Fig. E.1.

In order to deepen the knowledge about the employment and the development of Kano model, a recent state of the art analysis has been performed in [15].

References

1. Draft Federal Information Processing (1993) Standards publication 183 announcing the standard for integration definition for function modeling (IDEF0). Obtained through the internet: http://www.idef.com/pdf/idef0.pdf. Accessed 30 Nov 2011
2. Syque Quality (2011) The quality toolbox > IDEF. Obtained through the internet: http://syque.com/quality_tools/toolbook/IDEF0/idef0.htm. Accessed 30 Nov 2011
3. Kokolakis SA, Demopoulos AJ, Kiountouzis EA (2000) The use of business process modelling in information systems security analysis and design. Inf Manag Comput Secur 8(3):107–116
4. Yosuf KO, Smith NJ (1996) Modelling business processes in steel fabrication. Int J Proj Manag 14(6):367–371
5. Aguilar-Sauvén RS (2004) Business process modelling: review and framework. Int J Prod Econ 90(2):129–149

6. Pahl G, Beitz W (2007) Engineering design: a systematic approach, 3rd edn. Springer, London
7. Ogot M (2004) EMS models: adaptation of engineering design black-box models for use in TRIZ. In: Proceedings of ETRIA TRIZ future conference 2004, pp 333–345, Florence, 2–5 November 2004
8. Cox J, Goldratt EM (1996) The goal: a process of ongoing improvement. North River Press, Great Barrington
9. Dettmer WH (1997) Goldratt's theory of constraint: a system approach to continuous improvement. ASQ Quality Press, Milwaukee
10. Cascini G, Rissone P, Rotini F (2008) Business re-engineering through integration of methods and tools for process innovation. Proc Inst Mech Eng Part B J Eng Manuf 222(B12):1715–1728
11. Cascini G, Frillici FS, Jantschgi J, Kaikov I, Khomenko N (2009) Tetris—teaching TRIZ at schools. In: Cascini G (ed) EU lifelong learning program
12. Khomenko N (2010) General theory on powerful thinking (OTSM): digest of evolution, theoretical background, tools for practice and some domain of application. In: Japan TRIZ symposium 2010, Kanagawa Institute of Technology, Atsugi, 9–11 Sept 2010
13. Kano N, Seraku N, Takahashi F, Tsuji S (1984) Attractive quality and must-be quality. J Jpn Soc Qual Control 14(2):39–48
14. Borgianni Y, Cascini G, Rotini F (2010) Process value analysis for business process re-engineering. Proc Inst Mech Eng Part B J Eng Manuf 224(2):305–327
15. Mikulić J, Prebežac D (2011) A critical review of techniques for classifying quality attributes in the Kano model. Manag Serv Qual 21(1):46–66

Index